INTRODUCTORY MODAL LOGIC

Introductory Modal Logic

KENNETH KONYNDYK

University of Notre Dame Press
Notre Dame, Indiana 46556

Library of Congress Cataloging-in-Publication Data

Konyndyk, Kenneth.
  Introductory modal logic.

  Bibliography: p.
  1. Modality (Logic)   I. Title.
BC199.M6K66 1986   160   85-41007
ISBN 0-268-01159-1 (pbk.)

Manufactured in the United States of America

CONTENTS

PREFACE

This text is designed as an introduction to modal logic for a student who has studied elementary symbolic logic. It presents the symbolism, natural deduction proof procedures, and problems of modal logic, using the systems which have come to be known as T (or M), S4, and S5.

Since elementary symbolic logic texts differ with respect to symbolism, proof techniques, and rules, this book begins with a brief specification of symbols, rules, and proof technique to be used here.

Propositional modal logic and quantified modal logic are developed separately. The chapter treating propositional modal logic can be studied immediately after ordinary propositional logic. Although it introduces some metatheoretical considerations, it does not presuppose any prior acquaintance with metatheory. I consider some of the philosophical motivation for modal logic and do not try to pretend futilely that modal logic is or should be philosophically neutral, preferring instead to point out areas of philosophical controversy over modal logic.

Quantified modal logic raises even more philosophical questions. Which system is preferable depends quite directly on how certain metaphysical questions are answered. Indeed, the point of quantified modal logic is to represent the logical structure of certain correct metaphysical views. Accordingly, I develop the chapter on quantified modal logic so as to make clear where metaphysical choices have to be made, and I indicate what I think are the correct choices. There are plenty of references to discussions of the philosophical issues at stake. But I develop this chapter so a student can get a good grasp of quantified modal logic without being overwhelmed by the philosophical difficulties.

My presentation owes a great deal to the work of Saul Kripke and to the interpretations of Alvin Plantinga and Arthur N. Prior, but I am most deeply indebted to my colleagues Alvin Plantinga and

Thomas Jager. I am grateful to my colleagues in the Calvin College Philosophy Department for discussing this book in our weekly colloquium. They have suggested innumerable corrections and improvements. I also owe a great deal to my logic students who have served with varying degrees of willingness as the subjects of my experiments with this text.

I wish to express special thanks to Donna Kruithof, who exhibited patience and fortitude in typing and retyping the manuscript.

I began preliminary work on this project while I held an NEH Fellowship in Residence for College Teachers in 1975–1976, and I continued working on it during a sabbatical from Calvin College in the fall of 1977. I gratefully acknowledge the support of the NEH and of my home institution.

INTRODUCTION

Modal logic studies inferences that involve modalities of propositions. Here we will confine our attention to alethic modalities.

Propositions are true or false. However, we sometimes have occasion to note that a proposition is not merely true but that it is necessarily true, while other propositions, though false, are possibly true. So necessity and possibility are regarded as modes or modifications of truth (aletheia), and have come to be called alethic modalities.

Aristotle knew, and perhaps discovered, that these notions of *possibility* and *necessity* can each be expressed in terms of the other. He worked out some of this in *De Interpretatione*. Aristotle also did some of the first recorded work on modal inferences within the basic framework of the syllogism.

Like ordinary propositional logic, modal logic employs a notion of *implication* or *consequence*. In modal logic the notion of implication is strengthened by adding necessity to produce a concept labelled *necessary* (or "strict") *implication, necessary consequence,* or sometimes, *entailment.* This concept furnishes a more intuitive and accurate representation of many "if . . . , then . . ." sentences in ordinary usage, especially philosophical and theological usage, than classical propositional logic provides. One of the interesting things for us to note as we develop modal logic is which tautologies or theorems true of the classical implication relation ('⊃') have or lack counterparts for necessary implication.

Modal logic is usually developed as an extension of classical propositional logic and first-order quantification theory. That is the way it will be done here. Since most of the textbooks for learning first-order logic differ slightly from each other in symbolism, choice of rules, and formulation of rules and proof techniques, we will begin our treatment of modal logic by briefly specifying the formulation of propositional logic we will use. Once this is done, we turn to propositional modal logic, considering some of the motivation for it and different interpretations of it, as well as developing the formal apparatus. Then, in chapters 3 and 4, we do the same for quantified modal logic.

1

PROPOSITIONAL LOGIC

This book uses a natural deduction approach to validating modal inferences. We evaluate an inference by symbolizing it and deducing the symbolized conclusion from the symbolized premises, via simple and immediately obvious inference rules. For this approach we need rules for introducing and eliminating the various connectives in a proof. We will also use a rule of replacement, which enables us to replace equivalent expressions whenever we want to. We will consider an inference valid just in case there is a way of deriving the conclusion from the premises according to these rules.

The alternatives are the axiomatic and semantic approaches. In an axiomatic approach, we specify a set of axioms and simple rules for deriving theorems. We then evaluate an inference by extracting its logical form and determining whether or not that form is deducible from the axioms. This is a workable but a difficult and less intuitive way to proceed. The semantic approach is illustrated by truth tables in propositional logic. There we could evaluate an inference by extracting its logical form and then determining via a truth table, truth tree, or some other semantic device, whether or not that form was tautologous. These methods evaluate statement forms by assigning an interpretation, usually a truth value, to the variables, treating the connectives as mechanical ways of combining these values, and calculating the value of the statement form. This method can be very cumbersome and nonintuitive, but it has the virtue of providing us with a decisive method of showing argument forms valid and invalid. These semantic tests are examples of what are usually called *effective* procedures – they will produce a yes or no answer to our question about validity in a finite number of steps by a mechanical method.

Any one of these three approaches can be used to determine the validity of inferences in propositional logic. In choosing among them, therefore, we base our choice on factors like naturalness and ease of use. We use the natural deduction approach because the reasoning in symbols parallels the way we might argue in English for the validity of an inference. Usually it is the easiest way as well. Later on there will be occasion to look at axiomatizations and semantics.

We want a set of rules, then, that permit easy, straightforward, and fairly natural derivations. If simplicity and elegance were our only considerations, we would only need rules for introducing and eliminating our primitive operators. But simplicity and elegance take a back seat when our objectives are naturalness and convenience. Consequently some of our rules and equivalences will be redundant — we could get along without them.

We now review propositional logic in order to specify the symbolism and the rules to be used in this book.

The system of propositional logic presented here is the one found in most logic books. Different authors use different symbols or different rules, but each set of rules and each symbol system yields the same results when applied to an argument. This system of logic will be the base on which we build our modal logic.

What is presented here under the rubric of propositional logic is frequently labelled sentential logic or statement logic in other textbooks. Part of the reason is metaphysical controversy about what things are really the bearers of truth and falsity.[1] Here the word 'proposition' will ordinarily be used to refer to what are, properly speaking, the sorts of things that are true or false. This usage, however, is intended not to beg any metaphysical questions. Those who prefer 'sentence' or 'statement' should be able to substitute their choice *salve veritate.*

Since logic is customarily developed as a symbolic *language,* we will here refer to the symbolic representations of propositions as sentences of our symbolic logical language. These sentences will be spoken of as having truth values. Likewise, we will often refer to propositions (without trying to prejudice the issue of exactly what sorts of things they are) by means of sentences that are ordinarily used to express or assert them.

1.1 Symbolism

We use the following symbols for the operators:

SYMBOL	OPERATION	EXAMPLE	OTHER COMMONLY USED SYMBOLS
$\sim$	negation	$\sim A$	$\neg A, -A, \bar{A}$
$\supset$	implication	$A \supset B$	$A \rightarrow B$
&	conjunction	$A \,\&\, B$	$A \wedge B, A \cdot B, AB$
$\vee$	disjunction (or alternation)	$A \vee B$	
$\equiv$	material equivalence	$A \equiv B$	$A \leftrightarrow B$

Roman capitals from the beginning of the alphabet represent statements which are either true or false: $A, B, C, D. \ldots$ Lower-case letters from the latter part of the alphabet represent well-formed formulas: $p, q, r, \ldots$ Rules of inference and equivalences are stated in terms of p, q, r, etc., since the rules or equivalences apply to any formula exhibiting the structure under consideration.

We adopt the usual semantic, i.e., truth table, definitions of the operators or connectives.

p	$\sim p$
T	F
F	T

p , q	$p \vee q$
T T	T
T F	T
F T	T
F F	F

p , q	$p \,\&\, q$
T T	T
T F	F
F T	F
F F	F

p , q	$p \supset q$
T T	T
T F	F
F T	T
F F	T

p , q	$p \equiv q$
T T	T
T F	F
F T	F
F F	T

1.2   Rule of Replacement and Equivalences

The rule of replacement enables us to replace an expression with an equivalent expression. The expression being replaced may constitute either a part or the whole of a sentence or a well-formed formula. Such a replacement does not alter the truth value of the sentence within which it is made. The following is a list of familiar equivalences which we will use.

1.	De Morgan's	$\sim(p\,\&\,q) \equiv (\sim p \vee \sim q)$
	Theorems	$\sim(p \vee q) \equiv (\sim p\,\&\,\sim q)$
2.	Commutativity	$(p\,\&\,q) \equiv (q\,\&\,p)$
		$(p \vee q) \equiv (q \vee p)$
3.	Associativity	$(p\,\&\,(q\,\&\,r)) \equiv ((p\,\&\,q)\,\&\,r)$
		$(p \vee (q \vee r)) \equiv ((p \vee q) \vee r)$
4.	Distributivity	$(p\,\&\,(q \vee r)) \equiv ((p\,\&\,q) \vee (p\,\&\,r))$
		$(p \vee (q\,\&\,r)) \equiv ((p \vee q)\,\&\,(p \vee r))$
5.	Double Negation	$p \equiv \,\sim\sim p$
6.	Transposition	$(p \supset q) \equiv (\sim q \supset \sim p)$
7.	Material	$(p \supset q) \equiv (\sim p \vee q)$
	Implication	$(p \supset q) \equiv \,\sim(p\,\&\,\sim q)$
8.	Material	$(p \equiv q) \equiv ((p \supset q)\,\&\,(q \supset p))$
	Equivalence	$(p \equiv q) \equiv ((p\,\&\,q) \vee (\sim p\,\&\,\sim q))$
9.	Negation of Mate-	$\sim(p \equiv q) \equiv (\sim p \equiv q)$
	rial Equivalence	$\sim(p \equiv q) \equiv (p \equiv \sim q)$
10.	Exportation	$((p\,\&\,q) \supset r) \equiv (p \supset (q \supset r))$
11.	Absorption	$(p \supset q) \equiv (p \supset (p\,\&\,q))$
12.	Tautology	$p \equiv (p \vee p)$
		$p \equiv (p\,\&\,p)$

Obviously, this list could be extended if we so desired.

1.3   Rules of Inference

Rules of inference permit us to obtain new formulas – "new" in the sense that they need not be logically equivalent to any prior formula occurring in a proof. Unlike the rules for replacement of equivalences, rules of inference may *not* be applied within formulas or to parts of formulas.

Given the rule of replacement and our list of equivalences, the only rules of inference we require are *modus ponens* and implication introduction. However, as we have already said, elegance (using as little apparatus and as few rules as possible) is not our objective here.

We adopt the following rules because they correspond to basic and familiar inference patterns in everyday life, and they are the most convenient for us to use in our proofs:

1. Modus ponens
    (Implication elimination)
    $p \supset q$
    $\underline{p}$
    $\therefore q$

2. Modus tollens
    $p \supset q$
    $\underline{\sim q}$
    $\therefore \sim p$

3. Transitivity

$p \supset q$

$q \supset r$

$\therefore p \supset r$

4. Disjunctive introduction (Addition)

p

$\therefore p \vee q$

5. Disjunction elimination

$p \vee q$

$\sim p$

$\therefore q$

6. Conjunction introduction

p

q

$\therefore p \& q$

7. Conjunction elimination (Simplification)

$p \& q$

$\therefore p$

To these rules we add two more which will help simplify our proofs. These rules also have the virtue of being "natural" – we often use them in ordinary deductive reasoning. The first is the rule of *implication introduction*, often called "conditional proof." This rule provides a way of introducing implication statements into a proof.

8.   Implication introduction (Conditional proof)

p

.
.
.

q

.
.
.

r

.
.
.

s

.
.
.

$p \supset q$ impl intro

.
.

$p \supset r$ impl intro

.
.

$p \supset s$ impl intro

At any line in a proof, we may enter any formula as a line with the justification: assumption. When we enter such a line we indicate the assumption by drawing a horizontal line beneath it. We indicate the *scope* of the assumption by drawing a vertical line to its left, and continuing the line downward beside all subsequent lines until the assumption is discharged or terminated. An assumption is *discharged* or terminated by ending the vertical line and entering as the next line (or lines) of the proof a *conditional* (or conditionals), whose antecedent is the assumption and whose consequent is any line within the scope of the opening assumption that is not within the scope of any subsequent assumption. More than one conditional may be entered on the basis of a single assumption. The justification for such lines is *implication introduction*.

All the usual rules apply within the scope of an assumption, but no lines within the scope of a discharged or terminated assumption may be used to derive subsequent lines in a proof. No assumption may be discharged or terminated unless every assumption within its scope has been discharged or terminated.

An assumption also may be terminated by ending the vertical line indicating the scope of the assumption, entering no conclusions on the basis of the assumption, and proceeding with the proof as though the assumption had not been made. In such a case, we will say that the assumption has been terminated although it has not been discharged. This may be done when it becomes apparent that the assumption is not useful in proceeding with the proof.

Here are two ways of proving the same formula illustrating implication introduction:

$A \supset B$
$C \supset D$ $\therefore (A \& C) \supset (B \& D)$
| $A \& C$ asp
| A conj elim
| B MP
| C conj elim
| D MP
| $B \& D$ conj intro
$(A \& C) \supset (B \& D)$ impl intro

$A \supset B$
$C \supset D$ $\therefore (A \& C) \supset (B \& D)$
| A asp
| B MP
| | C asp
| | D MP
| | $B \& D$ conj intro
| $C \supset (B \& D)$ impl intro
$A \supset (C \supset (B \& D))$ impl intro
$(A \& C) \supset (B \& D)$ exp

Note that, by the statement of rule of impl intro, we could legitimately infer, e.g., $(A \& C) \supset C$, $(A \& C) \supset B$, etc., although in this problem there is no reason to do so.

This illustrates a use of impl intro within the scope of another usage of impl intro.

9. *Reductio ad absurdum* (Negation introduction)
This rule may be considered an abbreviation of a special application of implication introduction. We introduce it for convenience and because it parallels an ordinary and commonly used pattern of deductive inference. It is exactly like the rule of implication introduction, except that it gives different instructions for discharging the assumption. If within the scope of an assumption p, there occur as lines in the proof any formulas q and $\sim q$, the assumption may be discharged and $\sim p$ entered as the next line of the proof.

| p
| q
| $\sim q$
$\sim p$ RAA

Note than when a line within the scope of an assumption contradicts a line occurring outside, providing that the line outside the scope of the assumption does not occur within the scope of some other terminated or discharged assumption, equivalence (5) (double negation) enables us to bring the contradictory line within the scope of the assumption. Although "officially" required, this step may be eliminated in writing proofs.

The rule allows us to conclude that whatever implies a contradiction is false.

1.4   Proofs (or Derivations)

We have already been speaking of proofs, for the point of introducing these rules is the construction of proofs. A *proof* or derivation of a formula p is an ordered finite sequence of formulas, every member (or line) of which is either an assumption, a premise, or derived from previous lines by the rules, and whose last member (or line) is p, where this occurrence of p is not within the scope of any assumption.

An argument form is *valid* just in case there is a proof (or derivation) of its conclusion from its premises. And a formula p is a *theorem of propositional logic* if and only if there is a proof of p which uses no premises.

Example:

1.	$p \supset q$		assumption
2.		$q \supset r$	assumption
3.		$p \supset r$	1,2 Transitivity (Rule 3)
4.		$(q \supset r) \supset (p \supset r)$	impl intro
5.	$(p \supset q) \supset ((q \supset r) \supset (p \supset r))$		impl intro

Line 5 is a theorem.

1.5   Exercises

(There is no need for exercises of the usual sort here. Those given are related to remarks made in the text.)

1.   Show that the rule *modus tollens* is redundant (i.e., give a derivation of $\sim p$ from $p \supset q$ and $\sim q$ that does not use the rule *modus tollens*). Do the same for:

   rule 3–   Transitivity
   rule 4–   Disjunction elimination
   rule 9–   Reductio ad absurdum

2.   Show that the following formulas are theorems:
   (a) $(\sim p \supset \sim q) \supset ((\sim p \supset q) \supset p)$
   (b) $(p \supset (q \supset r)) \supset ((p \supset q) \supset (p \supset r))$
   (c) $(p \equiv (q \supset r)) \equiv (((p \& q) \supset r) \& ((q \supset r) \supset p))$
   (d) $p \supset (q \lor \sim q)$

3.   Note that among our rules of inference, we have rules for introducing and eliminating each of our logical connectives, except for negation. Why is there no negation elimination rule among our rules of inference?

2

PROPOSITIONAL MODAL LOGIC

Propositions, we have said, are either true or false. We have observed that there are propositions that, though true, could have been false, and they are distinct from those true propositions that could not have been false. Likewise, among false propositions there are those that could have been true and those that could not. Some truths are contingently true while others are necessarily true; some falsehoods are contingently false while others are necessarily false or impossible. Aristotle and the medievals thought of *necessity, possibility,* or *contingency* as modes or ways propositions may be related to truth or falsehood.

This list of modal notions fails to highlight an important subclass of necessary truths — those that are statements of entailment or necessary implication. In making assertions to the effect that one claim entails another, philosophers (and others as well) are claiming that some conditional is necessarily true. The antecedent entails the consequent; the statement that the antecedent implies the consequent is a necessary truth. Not all conditional sentences express this kind of conditional, as we shall see in the section on symbolization, but claims of this kind are common in philosophy and theology. Interest in the conditionals of this kind has provided a strong impetus to the development of modal logic in this century. Some of the basic arguments, however, apparently originated among the Stoics and were carried on by the Medievals.

Some grasp of these modal concepts is essential not only for understanding modal logic, but also for understanding the concept of *validity,* which is the central concept of logic. The usual definition of validity goes something like this: an argument form is valid just in case it is not *possible* for an argument of that form to have true premises and a false conclusion, or a form is valid if the conclusion *necessarily follows* from the premises. If we were to say instead that an argument form is valid if and only if there is (and has been) no argument of that form having true premises and a false conclusion, then we might be committed to accepting as valid an argument form that is patently invalid, just because no instance of it actually having true premises and a false conclusion has ever been presented or thought of.

Clearly the validity of an argument form must not depend solely on what arguments may or may not have been given in that form. The validity of an argument form is supposed to guarantee that we cannot use it to go from truth to falsehood. But then the modal term has been introduced into the definition, and we cannot get rid of it and still have what we want. Some primitive understanding of modal terms is essential to understanding the central concept of logic.

This chapter is concerned first with clarifying our grasp of the modal concepts and with the idea of a modal logic. The bulk of the chapter, however, is devoted to explaining three closely related systems of modal logic and developing facility in using them. The chapter concludes with remarks about some metaphysical assumptions implicit in our treatment and with some consideration of whether there is a "correct" system of modal logic.

2.1   The Modal Concepts

2.1.a   An Explanation of Necessity

Here we want to explain something of what we mean when we call propositions necessary, possible, or contingent. Since these notions are interdefinable, we concentrate on just one: necessity.

Most philosophers have thought there are some necessary truths. But there the widespread agreement ends. Which truths are necessary? Why are necessary truths necessary? How can necessary truths be characterized? The theories proceed in every direction.

Furthermore, there are different things one might legitimately mean by *necessity*. To call something a necessary truth is to imply that it somehow *has* to be true. But we sometimes say a claim is necessary when it is the consequence of certain antecedent conditions that are not within our (or anyone's) power to change. Perhaps *inevitable* is a more accurate word to describe this usage. But this usage is not exactly what we have in mind here, for it is possible that the antecedent conditions should have been different than they were, or that the laws relating these conditions to the claim under consideration could have been different. For example, that which has already happened in the past is often described as necessary in this sense of being inalterable, but the truth of a statement about a past event will not be necessary in the sense we have in mind. Likewise, physical and causal laws, sometimes said to be necessary, could be otherwise, and are therefore not necessary in the sense we are trying to describe. The same goes for such claims as

(1)   No human being can pole vault 1000 feet (under normal conditions with a standard pole, etc.).

Here we want to think of necessity as *logical necessity*, or as Alvin Plantinga describes it, "broadly logical necessity." Broadly logically necessary truths are those whose denials are self-inconsistent. Included in this group are truths of logic, for example, truths provable in first-order logic, such as,

(2)   For any statements p and q, if p is true and p implies q, then q is true.

and (3)   If all men are mortal and Socrates is a man, then Socrates is mortal.

So are truths of mathematics and set theory, such as

(4) $7 + 5 = 12$,

and (5)   The union of a set A with a set B is the same set as the union of B with A.

There is another group of truths, difficult to characterize, often labelled "analytic." Leibniz tries to characterize them by saying that the concept of the subject contains the concept of the predicate, e.g.,

(6)   All brothers are siblings,

and (7)   All bodies are extended.

Leibniz thinks that all necessary truths are analytic, but we will not make that rash claim here. We leave open the question of whether or not truths of logic and mathematics are analytic.

There is another broad class of necessary truths that is hard to characterize and whose members are sometimes difficult to pick out. They are not truths of logic or of mathematics, nor are they analytic. Yet their denials seem to lead directly to contradictions. Consider the assertion that some propositions are true. Its denial is

(8)   All propositions are false.

But that implies that proposition (8) is false. So if it is true, it is false, and of course if it is false, it is false. Therefore it is false, and its denial, some propositions are true, is true and necessarily so. Still other examples seem equally necessary, but their denials do not lead to such obvious arguments.

(9)   Nothing is both red all over and green all over at the same time;

(10) Nothing weighs more than itself;

(11) No numbers are human beings.

Still other candidates for this title of broadly logically necessary truth are hotly disputed, both as to their truth as well as to their necessity. Here are some examples:

(12) Man is a rational animal;

(13) There is no private language;

(14) For everything, there is a reason why it is so and not otherwise;

(15) Nothing happens without a cause;

(16) In an infinite stretch of time all possibilities (potentialities) are realized.

Although this concept of a logically necessary truth cannot be defined precisely and simply, the above examples help to communicate the concept. It is also helpful to look at some nearby concepts which should not be confused with it. Broadly logical necessity has already been distinguished from physical or causal necessity. But there are several more concepts to be distinguished from logical possibility. Once again our discussion draws heavily on Plantinga (*The Nature of Necessity*, chap. 1).

"Ungiveupable"

Our sense of necessity does not mean "ungiveupable." W. V. O. Quine, a leading contemporary philosopher, has urged philosophers to give up the analytic-synthetic (or necessary-contingent) distinction in favor of a notion of relative revisability. There are propositions that we are extremely reluctant to believe false, while we would quite readily reject others, given certain appropriate evidence. For example, a person may find a truth of logic almost impossible to give up, while she might be willing to give up the belief that there is a full cup of coffee on the desk, if she looks at it and it appears empty, even though she was sure she had just refilled it. But, Quine maintains, no propositions are in principle "ungiveupable," provided we are willing to adjust the rest of our beliefs.

Clearly we cannot equate relative "ungiveupability" with necessity, even though it may be a fairly reliable clue to necessary truth.
For a belief may be ungiveupable for reasons that have nothing to do
with its truth, and are unrelated to the necessity of its truth. A belief
may be a guiding belief for a person's life to such an extent that the
person may be psychologically incapable of giving it up. Or I may
find the belief that I exist ungiveupable, but its "ungiveupability"
has nothing to do with the necessity of its truth. Conversely a proposition may be a necessary truth, yet not have been thought of by anyone, and so not have achieved the status of being "ungiveupable."

"Unable to be rationally rejected"

Is a necessary truth one that cannot be rationally rejected? No,
for we can describe circumstances that make it clear that these concepts are distinct. I cannot rationally reject the proposition that I
am thinking, but it is not a necessary truth. And given that true
mathematical propositions are necessary, suppose that someone
whom I know to be a notorious liar, has just told me that Zorn's
Lemma is logically equivalent to the axiom of choice. Suppose further that I am totally ignorant of set theory and I know he has some
reason for wanting to deceive me on this occasion. Here I could rationally reject what is in fact a necessary truth.

"Self-evident"

Even if we had a sense of "self-evident" that did not also include
contingent propositions, we still could not say that all necessary
truths are self-evident. The example of complicated theorems of higher
mathematics suffices to show this. Could we say that a proposition
is necessary just in case it is either self-evident itself or deducible
from self-evident propositions by means of self-evident rules of inference? Consider Goldbach's conjecture — that every even number is the
sum of two primes. It is, if true, necessarily true, and if false, necessarily false. But which is it? There is no self-evident answer and none
is self-evidently deducible from self-evident truths, so far as we know
right now.

A priori

Is this hoary category the clue to picking out necessary truths?
One of the chief difficulties with this category of the *a priori* is under-

standing it. If we mean by it those truths that are known prior to sense experience, we can quickly answer our question No, since plenty of necessary truths are not known. Suppose then we say an *a priori* proposition is one knowable by someone independently of experience. Now we have a much more difficult question on our hands. Is God one of the persons whose knowing powers are under consideration here? If He is, then presumably all truths are *a priori*. Let us limit ourselves to human knowers. The philosopher Leibniz thinks that while God can know everything about every individual *a priori*, human knowers, due to the finitude of their knowledge, cannot. Human knowers, however, can know that they think and that they exist independently of sense experience. But these are not necessary truths.

All of these would-be explanations of necessity that we have rejected, except that of causal necessity, are epistemic notions. But the concept of necessity is an ontological one, not relative to anyone's actual knowledge or ability to know. Our illustrations have taken advantage of that fact.

2.1.b   *The Possible Worlds Picture*

The great modern rationalist Leibniz suggests another way of thinking of necessary truth. Leibniz introduces possible worlds as different ways God could have created the universe, or, to put it a little differently, possible worlds are alternative universes God could have created. Leibniz observes that God could have created different individuals from the ones that presently populate this world, he could have changed the natural laws, and he could have decided not to create at all. So a possible world is a total way things could be or could have been, and a necessary truth is a proposition that is true in every possible world.

Plantinga takes Leibniz's idea and tries to make it a bit more precise. He calls a possible world a *maximal* or *complete* possible state of affairs. He calls a possible state of affairs, S, *maximal* or *complete* just in case for every state of affairs, S either includes it or precludes it.

This Leibniz-Plantinga account does not provide an alternative way of explaining necessary truth, for it presupposes that anyone who uses it already understands the notion of *possibility* in the broadly logical sense. But if we have some grasp of that, we can understand what necessity in the broadly logical sense is, and we do not need to introduce worlds and maximal states of affairs. Of course

Plantinga does not give this account as an additional or independent way of explaining necessity, but as a help in understanding it.

If we have some idea of what is meant by calling a proposition a necessary truth, we can understand the other modal concepts by defining them in terms of necessity. A proposition is *possible* (or *possibly true*) just in case it is not necessary that it be false. Obviously then, a proposition is *impossible* just in case it is necessary that it be false. A *contingent* proposition is one which is both possibly true and possibly false. One proposition *entails* another if and only if it is not logically possible for the former to be true and the latter false. A pair of propositions is *compatible* or *consistent* if and only if their conjunction is possible, and *incompatible* or *inconsistent* just in case their conjunction is impossible. A pair is *contradictory* just in case both cannot be true and both cannot be false.

2.1.c   *A Map of the Modal Concepts*

Let us try to picture some of the relationships among these concepts. We begin by dividing propositions into two groups — true and false. Each of these groups can be divided again into necessary and contingent, yielding necessary truths and contingent truths, and necessary falsehoods and contingent falsehoods. The necessary falsehoods are the impossible propositions; the necessary truths together with the contingent truths and contingent falsehoods comprise the possible propositions. True propositions expressing entailments are necessary truths, although propositions standing in the relationship of entailment to each other need not be necessary, nor need they be true.

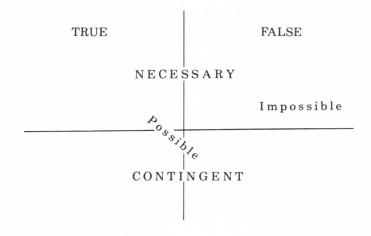

It may be helpful to have a few examples of the various categories:

> contingently true:
>> The average annual snowfall in Grand Rapids is 76 inches;
>> The Detroit Tigers won the 1984 World Series;
>
> contingently false:
>> The Houston Oilers won Super Bowl XVII;
>> The earth is not more than 40 million miles from the sun;
>
> impossible:
>> $7 + 5 = 13$;
>> There is a barber in Seville who shaves all and only those persons in Seville who do not shave themselves;
>
> entailment:
>> That I am wearing a blue sweater entails that I am wearing a sweater;
>> That I walk entails that I exist.

2.1.d Symbols and Definitions

We introduce the symbol $\Box$ as a sentence-forming operator on sentences. What falls within the scope of a $\Box$ must be a sentence of our logical language, and the result of prefacing a sentence with a $\Box$ is another sentence. $\Box p$ means that p expresses a necessary truth. $\Diamond$ is the possibility operator; $\Diamond p$ means that p expresses a possible truth.

Now we can express symbolically the definitions given at the end of 2.1.b:

p is possible	$\Diamond p = df \sim \Box \sim p$
p is necessary	$\Box p = df \sim \Diamond \sim p$
p is impossible	$\sim \Diamond p = df \ \Box \sim p$
p is contingent	$= df \ (\Diamond p \ \& \ \Diamond \sim p)$
p entails q	$p \rightarrow q = df \sim \Diamond (p \ \& \sim q)$
	$= df \ \Box (p \supset q)$
p and q are compatible	$= df \ \Diamond (p \ \& \ q)$
p and q are incompatible	$= df \ \sim \Diamond (p \ \& \ q)$
p and q are contradictory	$= df \ \sim \Diamond (p \ \& \ q) \ \& \sim \Diamond (\sim p \ \& \sim q)$
	or $(p \rightarrow \sim q) \ \& \ (\sim p \rightarrow q)$

Aristotle is the first philosopher we know of to discuss these modal relationships. In *De Interpretatione,* chapters 12 and 13, Aristotle struggles with the interrelationships among the modal notions and their negations. His account is marred by his failure always to distinguish clearly between possibility and contingency, but he eventually arrives at most of the interrelationships we have presented in our definitions and our previous discussion. For example, Aristotle discovered the equivalences (definitions)

$$\sim \Diamond p = df \ \Box \sim p,$$
$$\sim \Diamond \sim p = df \ \Box p,$$

and inferences

$$\Box p, \ /\therefore \Diamond p,$$
$$\Diamond p, \ /\therefore \sim \sim \Diamond p,$$
and $\qquad p, \ /\therefore \Diamond p.$

What interests us here is not how Aristotle knew these, nor whether he understood "possibility" and "necessity" in exactly the same way we use these terms, but rather that he saw logical relationships to be worked out.

2.1.e   Exercises (adapted from Aristotle, De Interpretatione, chap. 13)

A. Symbolize the following and show which groups contain statements that are all equivalent to each other:
1. It is possible that Tom is rational.
    It is not impossible that Tom is rational.
    It is not necessary that Tom is rational.
2. It is not possible that Tom is a mathematician.
    It is impossible that Tom is a mathematician.
    It is necessarily false that Tom is a mathematician.
3. It is possibly false that Tom is a cyclist.
    It is not impossible that it is false that Tom is a cyclist.
    It is not necessarily false that Tom is a cyclist.
4. It is not possibly false that Tom is bipedal.
    It is impossible that it be false that Tom is bipedal.
    It is not necessary that Tom is bipedal.

B. Give the contradictories of:
1. It is necessarily false that Tom is a mathematician.
2. It is impossible that it be false that Tom is bipedal.
3. It is not possible that Tom is a mathematician.

C. Are the following pairs compatible, incompatible, or contradictory?
   1. p is true.
      p is possibly true.
   2. p is true.
      p is possibly false.
   3. p is necessarily true.
      p is possibly true.
   4. p is necessarily true.
      p is contingently true.
   5. p is necessarily true.
      p is false.
   6. p is necessarily true.
      p is necessarily false.
   7. p is possibly true.
      p is possibly false.
   8. p is possibly true.
      p is necessarily false.

2.1.f Symbolizing Conditionals

The expressions "if... then...," "... implies...," and "... entails..." are often used, especially in philosophical and theological literature, in the sense we represent by our →. However, we should take note of certain important exceptions, so that we do not represent as modal arguments arguments that were never so intended by their authors. Sometimes conditionals may express causal relationships:

> If hydrogen and oxygen are mixed in the presence of a flame, then there will be an explosion;
> If you plant beans, then you will not get broccoli.

Sometimes they express subjunctive conditionals:

> If the Papacy had remained in Avignon, then today's Pope would be a Frenchman;
> If Bach had written the *1812 Overture,* he would have known the melody of the *Marseillaise.*

And there are other kinds of conditionals:

> If Jager buys a bicycle, it will be either a Schwinn or a Fuji;
> If you sit in the front of the plane, then you may not smoke;

> If you took the $100 from Jones' drawer, then you have violated the law;
> If you expect your share of the inheritance, then you must stop your carousing.

None of these expresses a conditional that is a necessary truth; in each case it is possible for the antecedent to be true and the consequent to be false. Yet each of these may, in the proper context, express a true conditional.

2.1.g   The Necessity of the Consequence and the Necessity of the Consequent

During the Medieval period, the ability to distinguish the necessity of the *consequence* from the necessity of the *consequent* became a required item in the philosopher's repertoire. St. Thomas Aquinas makes a typical use of this distinction to resolve a puzzle about God's foreknowledge and human free will in his *Summa Contra Gentiles*, part I, chapter 67. But even St. Augustine and Boethius show an awareness of this distinction, though less clear than St. Thomas's, in their treatments of the same puzzle many centuries earlier.[2]

Basically, the distinction is between two different ways of taking the scope of necessity in a conditional. The occurrence of the word 'necessarily' in a conditional may signal the necessity of the connection between the antecedent and the consequent of that conditional, that is to say, the necessity of the *consequence*. Or 'necessarily' may indicate the necessity of the *consequent* of the conditional. Representing the situation symbolically, we are distinguishing

$$\Box \ (p \supset q) \quad \text{necessity of the consequence}$$

from

$$(p \supset \Box q) \quad \text{necessity of the consequent.}$$

The symbolization above makes it immediately clear that there is a distinction to be made and makes clear what the distinction comes to. Many earlier writers seem to think there are two different kinds of necessity involved. Once we see the distinction represented symbolically, it may become difficult for us to see how astute philosophers have managed to confuse these two readings.

This confusion is often engendered by the ambiguous ways of expressing necessity in conditionals in natural languages. For example, consider the following claim put in the mouth of Evodius by St. Augustine in the latter's *On the Free Choice of the Will*:

"Since God foreknew that man would sin, that which God foreknew must necessarily come to pass."

What is the scope of "necessarily" here? Is Evodius claiming that

> If God foreknew that man would sin, then it *follows necessarily* that man would sin (or, equivalently, It is necessary that if God foreknew that man would sin, then man would sin),

or that

> If God foreknew that man would sin, then it is (a) necessary (truth) that man would sin?

The philosophical question being discussed in the passage is whether God's foreknowledge necessitates what he knows. Which reading we choose could make all the difference in how we answer that question. However, since the question of God's knowledge gives rise to additional complications and confusions, it might be better to pursue the question of interpretation with the help of a different example.

Most philosophers have held that it is impossible to know, actually know, something false. In other words, if I know p, it necessarily follows that p is true, i.e., $\Box(S$ knows $p \supset p$ is true). I might, on some occasion, express this truth by saying that if I know there is coffee in my cup, then necessarily there is coffee in my cup. However, this way of expressing myself might lead someone to think what I said expresses the necessity of the consequent:

> ((I know there's coffee in my cup) $\supset \Box$ (there's coffee in my cup)).

But a brief moment's reflection is enough to lead us to reject that as a serious interpretation, assuming that I was trying to say something true. For if this reading were correct, I would generate a quick but specious argument for the necessary truth of everything I know or anyone knows.

In spite of what may now seem obvious, there is a strong and recurring temptation in philosophy and theology to transfer the necessity of the connection in a conditional to the consequent of the conditional. It is important to resist this temptation. Indeed, it is a good rule of thumb in translating from natural language into symbols to suppose that necessity expressed in a conditional should be taken as expressing necessity of the *consequence,* unless there is overwhelming specific evidence to the contrary.

Remember that to assert the necessity of the consequence is to assert that it is broadly logically impossible for the antecedent of the conditional to be true while the consequence is false. Asserting the necessity of the consequent, on the other hand, is to claim that the antecedent implies the necessary truth of the consequent.

2.1.h Exercises

A. Which of the following express necessary conditionals?
   1. If you study the history of music, then you cannot omit Bach.
   2. If Gacy killed those young men intentionally, then he is guilty of murder.
   3. If a triangle is equilateral, then it is necessarily equiangular.
   4. If Hitler had conquered Russia, then Khrushchev would not have come to power.
   5. If the millage does not pass, the Superintendent will have to resign.
B. Symbolize the following:
   1. If God does not exist, then he cannot come into existence.
   2. If God does not exist, then his existence is impossible.
   3. If God exists, his existence is necessary.

2.2 The Logic of Modality

2.2.a The Need for a Logic of Modality

Earlier we said that modal logic was the logic of these notions of possibility and necessity. Now that the concepts are before us, what should a *logic* of these concepts be like and what should it tell us?

Clearly, a *modal logic* should provide us with a way of exhibiting the logical structure of those inferences that use modal concepts in a way which affects their validity. There is no need to introduce more apparatus than we have in ordinary propositional logic to validate inferences of the form

$$\Box p \supset \Box q \qquad \Diamond p \supset \Diamond q$$
$$\Box p \qquad \text{or} \qquad \Diamond p$$
$$\therefore \Box q \qquad \therefore \Diamond q$$

since both are simply instances of modus ponens.

There are, however, some inferences in which these modal concepts obviously play an important role, and the validity of those inferences cannot be shown in ordinary propositional logic. Consider the inference

> It is necessary that all black cats are black
> ∴(It is true that) All black cats are black,

and its counterpart for possibility,

> (It is true that) Jager is a bicyclist
> ∴It is possible that Jager is a bicyclist.

These inferences are obviously valid. Yet we cannot show that they are, given only the resources of propositional logic. In each case we would have to represent the structure of the argument as

$$p$$
$$\therefore q$$

which is an invalid inference.

Although the above way of symbolizing the inference in propositional logic fails to show its validity, perhaps there is some other way to represent this inference in terms of truth-functional connectives alone that does show the validity of this inference. Can $\Box p$ (i.e., p is a necessary truth) be represented truth functionally? Our set of connectives is functionally complete, that is, they are adequate to represent all truth functions.

If $\Box p$ can be adequately represented truth functionally, there will have to be some sentence using only p and the usual truth-functional connectives that is equivalent to $\Box p$ and that makes valid the patently valid inferences involving $\Box p$. There are only four possible ways to express $\Box p$ truth functionally, represented here by the four truth functions, f_1–f_4.

TRUTH FUNCTIONS

p	$f_1(p)$	$f_2(p)$	$f_3(p)$	$f_4(p)$
T	T	F	T	F
F	T	F	F	T

1. Suggestion f_1 is a truth-function that makes $\Box p$ true no matter what p is. Hence it makes $\Box p$ equivalent to $(p \lor \sim p)$;
2. Suggestion f_2 is a truth-function that makes $\Box p$ false no matter what p is. Hence it makes $\Box p$ equivalent to $(p \mathrel{\&} \sim p)$;

3. Suggestion f_3 is a truth-function that makes $\Box p$ have the same truth value as p. Hence it makes $\Box p$ equivalent to p;

4. Suggestion f_4 is a truth-function that makes $\Box p$ have the opposite truth value of p. Hence it makes $\Box p$ equivalent to $\sim p$.

Obviously each of these is unsatisfactory as a representation of $\Box p$.

1. Suggestion f_1 fails to make valid $\Box p \supset p$, which is surely valid. $(p \vee \sim p) \supset p$, for example, is obviously invalid.

2. Suggestion f_2 makes valid $\Box p \supset p$, because $(p \& \sim p) \supset p$ is valid. But it also makes valid the obviously invalid $\Box p \supset \sim p$, incorrectly making it equivalent to $(p \& \sim p) \supset \sim p$, a valid formula.

3. Suggestion f_3 makes $\Box p$ equivalent to p; while this validates $\Box p \supset p$, it also would validate $p \supset \Box p$, which is patently invalid.

4. Suggestion f_4 makes $\Box p$ equivalent to $\sim p$. This fails to make $\Box p \supset p$ valid.

The failure of all four possible suggestions shows that the logic of necessity (of our $\Box$) is not truth functional. Valid modal inferences cannot be systematized using only truth-functional propositional logic. Some additional rules for modal inferences will have to be developed.

2.2.b Logical Relationships

We mentioned earlier that Aristotle saw that statements involving modal concepts had logical relationships to each other, and that he began to work this out. One of the things a modal logic should provide is an account, perhaps in the form of definitions, of the direct relationships between the modal concepts.

The medievals spelled out these relationships in their squares of opposition and rules of equipollence (something like logical equivalence). On the following page is a loosely translated presentation of a scheme that was standard by the twelfth century.[3] Note that contingency is still not distinguished from possibility here. Each of the groups of sentences beside the "corners" of the square consists of equipollent or equivalent expressions. The groups at the ends of the diagonals are contradictory, the upper groups are contraries (they cannot both be true), the lower groups are subcontraries (they cannot both be false), while the sides represent subalternation (the upper group on a side implies the lower). The reader can easily verify that this works out, except for contingency.

Modern work in modal logic, and much of the modal logic in this book, owes its beginning to the work of C. I. Lewis. Lewis was unhappy with material or logical deducibility. Russell and Whitehead, when they introduced the '⊃' in *Principia Mathematica,* called the relationship *"material implication"* and this connective the "material conditional" precisely because they recognized it differed from what is often called implication. The main problem with it is that it seems to allow too much: propositions turn out to "materially imply" lots of propositions we would not want to say that they "imply." This objection can be summarized formally in the "paradoxes of material implication":

$p \supset (q \supset p)$, i.e., a true proposition is implied by any proposition

and $\qquad \sim p \supset (p \supset q)$, i.e., a false proposition implies any proposition.

That this understanding of implication led to these paradoxes was not something Lewis discovered. The ancient Megaric philosopher

It is not possibly not the case
It is not contingently not the case
It is impossibly not the case
It is necessarily the case

It is not possibly the case
It is not contingently the case
It is impossibly the case
It is necessarily not the case

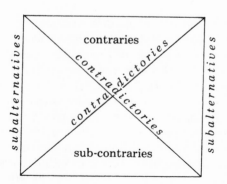

It is possibly the case
It is contingently the case
It is not impossibly the case
It is not necessarily not the case

It is possibly not the case
It is contingently not the case
It is not impossibly not the case
It is not necessarily the case.

Philo proposed defining the conditional in just the way material implication is defined – false when the antecedent is true and the consequent false, and true otherwise. This was quickly criticized by his fellow philosophers, including his teacher Diodorus Cronos. Furthermore, these ancients discussed the understanding of the conditional that Lewis favors – that a conditional is true precisely when the negation of the consequent is incompatible with the antecedent.[4]

Lewis calls his definition of implication, which we adopt, "strict implication"; it is also referred to as "entailment":

$$p \rightarrow q \text{ iff } \sim \Diamond (p \,\&\, \sim q), \text{ which is equivalent to } \Box(p \supset q).$$

However, this sense of implication is not without apparent paradoxes of its own. Although the paradoxes of material implication do not hold for it, there are analogues of those paradoxes:

$\Box p \supset (q \supset p)$	a necessary truth is strictly implied by any proposition;
$\sim \Diamond p \supset (p \rightarrow q)$	an impossible proposition strictly implies any proposition.

Lewis intended his sense of strict implication to represent the relationship of logical deducibility. That is, if p entails q, q should follow logically from p. Yet, as one can see from the analogues to the paradoxes of material implication, Lewis's strict implication is not completely faithful to the relationship of logical deducibility. It permits the deduction of a necessary truth, such as $2 + 2 = 4$, from any proposition.

A. R. Anderson and Nuel Belnap, Jr., have criticized Lewis's notion of strict implication, while concurring with his objection to material implication. They claim that in addition to the requirement that there be a necessary connection between a proposition and what it implies, the premises or antecedent must be relevant to what is implied. As we have just seen, the paradoxes of strict implication, like the paradoxes of material implication, expose the possibility of having a conclusion strictly implied by irrelevant premises. Anderson and Belnap have done extensive work on a system of logic designed to remedy these defects.[5]

In the face of these objections, why do we adopt the Lewis account of entailment? Why do we accept these paradoxes?

The "paradoxes," as Lewis likes to point out, are the consequences of straightforward and obviously acceptable rules. Thus our options are either to accept the paradoxes or to give up some obviously correct rule. Consider this demonstration of the claim that from a contradiction anything follows.

$$A \& \sim A \qquad \text{prem}$$
$$A \qquad \text{conj elim}$$
$$A \vee B \qquad \text{disj intro}$$
$$\sim A \qquad \text{conj elim}$$
$$\therefore B \qquad \text{disj elim}$$

One of the rules used in this proof will have to be declared incorrect and forbidden in a logic of "logical deducibility," if Anderson and Belnap are right. Which rule should we give up? Which rule is such that its conclusion is not "logically deducible" from its premise or premises?

We could make the point in a slightly different way. It is plausible to regard the following as expressing properties of the entailment relationship and therefore theorems of any adequate system representing entailment:

1. $((p \rightarrow q) \& (q \rightarrow r)) \rightarrow (p \rightarrow r)$ transitivity
2. $(p \& q) \rightarrow ((p \vee r) \& q)$ additivity
3. $((p \vee q) \& \sim p) \rightarrow q$ disjoinability

But any system that has the above as theorems (and uses modus ponens and conjunction as rules of inference) yields the alleged paradox as a theorem:

1′. $(p \& \sim p) \rightarrow ((p \vee q) \& \sim p)$ substitution instance of (2) above
2′. $((p \vee q) \& \sim p) \rightarrow q$ (3) above
3′. $(((p \& \sim p) \rightarrow ((p \vee q) \& \sim p))$ substitution instance of (1)
     $\& (((p \vee q) \& \sim p) \rightarrow q)) \rightarrow$ above
     $((p \& \sim p) \rightarrow q)$
4′. $(p \& \sim p) \rightarrow ((p \vee q) \& \sim p)$ 1′, 2′, conjunction
     $\& (((p \vee q) \& \sim p) \rightarrow q)$
5′. $(p \& \sim p) \rightarrow q$ steps 2′, 4′ MP

The price of eliminating the paradoxes of strict implication is high: we must give up a rule and the corresponding theorem from the proofs just given. Lewis found it less strange to accept the paradoxical formulas than to reject any of these rules or theorems. Anderson and Belnap find the formulas so paradoxical that they choose to give up a rule and corresponding theorem.

They argue that the rules of disjunction introduction and disjunction elimination, and the theorems labelled "additivity" and "disjoinability" are invalid as they stand. The theorems involve a logically disreputable and unfit sense of disjunction, if one's aim is (as Lewis's is) to come up with a calculus of entailment that accurately represents the properties of logical deducibility. In effect their criticism goes like this: although the rules in question are valid for the truth-functional 'v' introduced in the classical propositional calculus, they permit fallacies of *relevance*. In any respectable sense of deducibility the premises have to be relevant to the conclusion, but the rules in question permit us to deduce conclusions from premises which are patently irrelevant. E.g.,

> Either Bach wrote the Coffee Cantata or there is a largest
>   prime number;
> There is no largest prime number;
> Therefore, Bach wrote the Coffee Cantata.

If we introduce restricted versions of these rules, restrictions that forbid the deduction of irrelevancies, then we will find that we can no longer deduce q from $(p \,\&\, {\sim}p)$.

Here we favor Lewis's conclusion, retaining the rules Anderson and Belnap reject and accepting the so-called paradoxes of strict implication. Two of the three modal systems presented here are ones formalized by Lewis, though none is the system Lewis himself favors. Lewis refuses to adopt "officially" his stronger systems S4 and S5, developed later in this chapter, because they include as a theorem

$$(p \rightarrow q) \rightarrow ((q \rightarrow r) \rightarrow (p \rightarrow r)).$$

Lewis's favored system, S2, includes transitivity,

$$((p \rightarrow q) \,\&\, (q \rightarrow r)) \rightarrow (p \rightarrow r),$$

but not the "exported" form above. Lewis is unable to believe that $(q \rightarrow r) \rightarrow (p \rightarrow r)$ is logically deducible from $p \rightarrow q$.

In moving from truth-functional propositional logic to modal logic, we leave behind the genuinely puzzling paradoxes of material implication and obtain a system that better represents entailment or logical deducibility. Our goal, however, is not to represent strictly by means of a formal system what "implies" or "is logically deducible from" means in English. Rather it is to represent the logic of necessity, where necessity is interpreted as "broadly logical necessity."

We thus elect to use a system that sometimes fails to preserve relevance. However, we may be surprised at the relevance a clever person can find. Russell is reputed to have been challenged to prove that the necessarily false hypothesis $2 + 2 = 5$ implied that he was identical with the Pope. Russell replied, "You admit that $2 + 2 = 5$, but I can prove that $2 + 2 = 4$; therefore $5 = 4$. Taking 2 away from both sides, we have $3 = 2$; taking 1, $2 = 1$. But you will admit that I and the Pope are 2. Therefore, I and the Pope are one. Q.E.D."

2.2.c   Systems of Propositional Modal Logic

As I already have suggested, there are many different systems of modal logic. These are different not merely in the sense that they have different rules or axioms. For different axiom sets can still yield the same system (i.e., the same set of theorems) and different sets of rules can still validate exactly the same inferences. Here we mean that there are different systems in the sense of having different theorems and validating different inferences.

Which system a philosopher chooses depends on his philosophical commitments and his interpretation of the modal operators. Well then, is there not one correct system of modal logic? As is often the case, the answer is Yes and No; Yes in one sense and No in another. To see why, we must look at the question more closely.

There are different ways to consider a logical system. We can think about it as a system of symbols for which we give rules. We have rules for picking out certain approved strings of symbols and rules for manipulating these strings. If we work in this fashion, we can study a system without any reference to its intended use. There may be several different uses for a system, obtained by interpreting the symbols of the system in different ways. For example, what you learned as the calculus of propositions can also be interpreted as a calculus of classes.

As an exercise, look at this interpretation. The letters now denote classes: ' ~ ' becomes complementation, '⁻'; '&' becomes intersection, '∩'; disjunction 'v' becomes union, '∪'; '⊃' becomes inclusion, '⊂'; and '≡' becomes equality of classes, '='. Look back at the equivalences in the first chapter and rewrite them as statements about classes, paying attention to what they say under this new interpretation.

If we are speaking of a formal system and want to ask whether it is a correct system in a more than merely formal sense, we must know what the intended interpretation is – what the system is supposed to represent. Then a system can be said to represent that correctly or incorrectly. We would determine this by looking at the theorems (interpreted, of course) and considering whether or not they are true, and whether all the truths we are interested in are theorems.

We have just seen that systems may have several interpretations. The uninterpreted formal system for the propositional calculus provides a correct system under more than one interpretation. But it can be incorrect under other interpretations. When we ask about the correctness of a formal logical system and, in the present case, a modal system, we must have in mind some intended interpretation of the formal system.

This question is complicated in the case of modal logic because there are more than ten formal systems which have been proposed and studied, and there are several interpretations of the modalities. Here we have presented an interpretation of modalities; our primitive (undefined within the system) modal operator represents "broadly logical necessity." Is there a system that is correct under that interpretation? Our tentative answer is that there is one and that it is the third of the three presented later in this chapter. For a different sense of "necessity," one of the other systems might be correct.

Why present three systems here? Partly because of the way they fit together. The first one, T, is built on and thus contains propositional logic. The second, S4, contains T and more. S5 contains S4 and more. Thus, if S5 is correct, none of these systems will make valid any invalid inferences. But the latter systems validate more inferences. As we go along, we will point out the differences and their philosophical significance. Our view is that S5 is superior (more correct) because it includes as theorems truths which the others lack. More conservative philosophers may choose one of the weaker systems. It is also very easy to give natural deduction rules for these three systems. We can change from a weaker system to a stronger one (more theorems) with a simple modification of one rule.

2.3   The System T

This system is the only one of the three we study which was not proposed by C. I. Lewis. It fits so neatly and naturally into the "family" of systems proposed by Lewis that it is regarded as a Lewis modal system.[6] A version of it was studied by von Wright in 1951 and called M. However, it had been proposed earlier in a different form by Robert Feys, who called it T. These two forms were subsequently shown to be equivalent. Our presentation will differ from both of theirs.

First, recall the symbols already introduced:

1. □, an operator attaching to sentences and forming sentences, read as "it is logically necessary that," or "necessarily." This is attachable to any well-formed formula or sentence of propositional logic.
2. ◊, the so-called weak modal operator, also attaching to sentences and read as "it is logically possible that," or "possibly."
3. →, representing a logically necessary conditional, and read as "necessarily if . . . , then . . . ," or also as "entails" or "(strictly) implies."

We will not introduce another symbol for logical contingency. When we want to indicate that a proposition, p, is contingent, we will write

$$\Diamond p \ \& \ \Diamond{\sim}p.$$

The rule of replacement of equivalents is retained and the following definitional equivalences added:

$$\Box p \leftrightarrow {\sim}\Diamond{\sim}p$$
$$\Diamond p \leftrightarrow {\sim}\Box{\sim}p$$
$$(p \rightarrow q) \leftrightarrow {\sim}\Diamond(p \ \& \ {\sim}q)$$
$$(p \rightarrow q) \leftrightarrow \Box(p \supset q)$$
$$(p \leftrightarrow q) \leftrightarrow ((p \rightarrow q) \ \& \ (q \rightarrow p))$$

These formulas may replace each other wherever they occur, with the justification *definition* (def).

Exercises: substitute equivalents until both sides are the same:

1. $\Box((p \ \& \ q) \supset r) \equiv {\sim}\Diamond(q \ \& \ (p \ \& \ {\sim}r))$
2. ${\sim}\Diamond(\Box p \ \& \ {\sim}\Box q) \equiv ({\sim}\Diamond q \rightarrow \Diamond{\sim}p)$
3. $\Box(p \ \& \ (q \lor r)) \equiv (({\sim}p \lor {\sim}q) \rightarrow (p \ \& \ r))$

2.3.a   Rules

What are the right rules for modal logic? How should you proceed if you had to devise rules by yourself? Perhaps the best thing to do is to look carefully at the concept you want to represent, trying to discover on an intuitive level what its implications are. Next you would propose some formal rules grounded in what seem to be basic inference patterns and investigate their consequences.

When we think about broadly logical necessity this way, one obvious implication, so obvious we might overlook it, is that whatever is logically necessary is true. We represent that in our symbolism like this:

(1) $\Box p \supset p.$

Another characteristic of logically necessary propositions is that any proposition that necessarily follows from a logically necessary proposition is itself logically necessary. We may represent that as follows:

(2) $(\Box p \ \& \ \Box(p \supset q)) \supset \Box q.$

In writings on modal logic this is usually put symbolically in the equivalent form

(3) $\Box(p \supset q) \supset (\Box p \supset \Box q).$

Furthermore, if we reflect on what it means to be a truth-table tautology (or theorem of ordinary propositional logic) it seems clear that all tautologies are necessary truths. Hence, we may adopt as a rule

(R)   If p is a theorem of propositional logic or a tautology, then $\Box p$ is a theorem of modal logic. ($\vdash R$, $/\therefore$ $\vdash \Box R$).

As the clever reader no doubt suspects, we are developing the system T of modal logic. Indeed we have developed it. Any set of axioms and rules for the propositional calculus with (1) and (3) added as axioms and (R) added as a rule constitutes an axiomatization of T. Thus we can see that T is a comparatively fundamental system of modal logic, inasmuch as it is based on some universal but fundamental observations about the logical behavior of the concept of broadly logical necessity.

Our aim in this book is to present natural deduction rules for the systems we consider, so let us turn to that task. Since we have introduced $\Box$ as our primitive (undefined) operator, we should give a rule for introducing it and a rule for eliminating it. These rules will be counterparts of (1) and (3), which represent our observations about the behavior of the concept of broadly logical necessity.

The necessity elimination rule

The rule of necessity elimination (*nec elim*) is straightforward and obvious, following (1):

$\Box p$
$\therefore p$ (nec elim)

Stating a rule of necessity introduction will take a little more machinery. The machinery is not absolutely necessary, but once we have it, it will facilitate giving proofs and make it easier to switch to the other systems of propositional modal logic. We might take our cue from (2) above and use as our rule

$\Box(p \supset q)$
$\underline{\Box p}$
$\therefore \Box q$

But this rule would not be sufficient without a way of introducing the theorems of propositional logic as necessary truths. That is, we would use a rule like (R). And we would need to know and recall a wide range of these tautologies or theorems.

Necessity introduction subproofs

Instead, we will provide a process of deducing necessarily true conclusions from necessarily true premises by introducing *necessity introduction subproofs* and providing *reiteration* rules, to be specified shortly. The subproof rule used here is adapted from Frederick B. Fitch, *Symbolic Logic* (Ronald Press), 1952.

A *necessity introduction subproof* is a subordinate proof, analogous to the implication introduction rule. Into this subproof, however, we may bring only lines that are necessary, i.e., lines of the form □p. That way we are assured that anything we deduce within such a subproof is also *necessary*. The guiding insight is that whatever follows from what is necessary is itself necessary. If we permit only formulas that are necessary to be brought into such a subproof, then we may conclude that anything we deduce from such formulas within the subproof is likewise necessary. When we terminate such a subproof, we may write the last line or any other line within the subproof (provided it is not within the scope of a further subproof within the necessity introduction subproof) outside the subproof and prefix it with the □. We are permitted to bring only lines of the form □p into a necessity introduction subproof, a process we call reiteration. In T, when we enter such a line within the subproof, we remove the □.

The subproof rule thus reflects the idea that anything deduced from premises that are necessary is itself necessary.

The necessity introduction subproof rule

A nec intro subproof is opened by entering a formula, *p*, as a line, provided that either

(i)   entering *p* is justified by the appropriate rule of reiteration;

or (ii)   *p* is the assumption for an impl intro subproof within the scope of the nec intro subproof. (It may be helpful to recall that reductio ad absurdum proofs are cases of impl intro subproofs.)

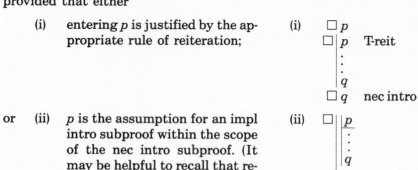

When a nec intro subproof is opened, we will draw a vertical line with a □ to the left of it alongside the subproof. The line is continued until the subproof is discharged and indicates the scope of the subproof, as the above schematic examples illustrate. Any subsequent line within a nec intro subproof must be either

	(i)	the result of using the appropriate reiteration rule;
or	(ii)	an assumption for an impl intro subproof within the scope of the nec intro subproof;
or	(iii)	the result of applying our rules to lines within the scope of the nec intro subproof;
or	(iv)	the opening line of another nec intro subproof within the scope of the original.

The necessity introduction rule

A nec intro subproof may be terminated at any point, and any line, p, occurring within the subproof and which is not within the scope of an undischarged assumption within the subproof may be entered as a line prefixed by a □, with the justification "nec intro."

$$\begin{array}{ll} \Box & \vdots \\ & \vdots \\ & p \\ & \vdots \\ & \vdots \\ & q \\ \Box\, q & \text{nec intro} \end{array}$$

The termination of the nec intro subproof is indicated by discontinuing the vertical line marking the scope of the subproof.

The complicated-sounding restrictions are due to our desire to maintain maximum flexibility while avoiding invalid inferences. This is the only rule of the propositional modal system T that may take some effort to learn. But that effort will pay off in the great ease with which you can learn S4 and S5. To obtain these systems, we merely need to modify our reiteration rule.

We can best see how to use the nec intro rule and the companion subproof rule by considering an example:

Prove $((p \to q) \& \sim q) \to \sim p$

1.	$\Box$ $(p \to q) \& \sim q$	impl intro asp
2.	$p \to q$	& elim
3.	$\Box(p \supset q)$	def
4.	$p \supset q$	nec elim
5.	$\sim q$	& elim
6.	$\sim p$	MT
7.	$((p \to q) \& \sim q) \supset \sim p$	impl intro
8.	$\Box[((p \to q) \& \sim q) \supset \sim p]$	nec intro
9.	$((p \to q) \& \sim q) \to \sim p$	def

The T-reiteration rule

As our statement of the necessity introduction subproof rule makes clear, the only way to bring "outside" formulas into the subproof without making assumptions is by the rule of reiteration. Now we formally introduce such a rule for T.

> Where $\Box p$ occurs as a previous line in a proof, p may be entered as a line within a nec intro subproof with the justification "T-reit," unless that occurrence of $\Box p$ lies within the scope of a discharged assumption or within the scope of a terminated nec intro subproof.

Notice that we may reiterate only formulas of the form, $\Box p$, and when we do, we must remove the $\Box$. It may seem that what is happening is more like nec elim than reiteration, but after we consider S4 and S5, it will be clearer why we choose to call it "reiteration." For the moment, it may be sufficient to observe that when we take formulas "out" of the nec intro subproof, we do put the $\Box$ back on them. Thus it gives us a way to introduce the $\Box$(necessity) onto formulas that did not have $\Box$'s on them earlier in the proof.

Why do we remove the $\Box$? Intuitively, what we are doing in a nec intro subproof where T-reit is the only allowable reiteration rule is to bring a sentence that is necessary into the subproof, deduce consequences from it, and conclude that those consequences are necessary. If we did not remove the $\Box$, we would in effect be deducing consequences from $\Box p$ rather than from p. Furthermore, the essence of T is that we do not assume that if p is necessary, then it is necessary that p is necessary ($\Box p \supset \Box\Box p$). But without the stipulation that when we T-reiterate we must remove the $\Box$, this formula would be easy to prove.

Here is an example illustrating the use of necessity introduction subproofs and T-reiteration.

1.	$p \to q$	prem
2.	$\Box p$	prem/ $\therefore \Box q$
3.	$\Box(p \supset q)$	1, def
4.	$\Box \mid p \supset q$	3, T-reit
5.	$\mid p$	2, T-reit
6.	$\mid q$	4,5 MP
7.	$\Box q$	6, nec intro

2.3.b   Hints

Here are a few hints that will be quite helpful in using these rules:

(1)   Remember that *in T,* the only way to bring something into a nec intro subproof is by means of T-reit. Any line brought in must be of the form $\Box p,$ and the $\Box$ must be removed.

(2)   Since any line and only lines of the form $\Box p$ may be entered with the use of T-reit, keep in mind that

$$(p \rightarrow q) \leftrightarrow \Box (p \supset q)$$
$$\sim \Diamond p \leftrightarrow \Box \sim p.$$

(3)   You may use impl intro within a nec intro subproof. See the earlier example. As the rule is stated, it does not require that you discharge any assumption introduced within a nec intro subproof. If you make an assumption and then see that it does not lead to anything useful, you may simply terminate it, and the proof you produce will still be "official." Remember, though, that you may not apply nec intro to a formula from an interior subproof unless it is first removed from the scope of the assumption.

(4)   Observe that by using impl intro as illustrated in the earlier example, you can prove formulas of the form $\Box p$ without being given any premises. All formulas derivable from our rules without the help of premises are *theorems.*

(5)   Typically, you will use the rule nec intro by constructing a nec intro subproof and applying nec intro to its last line. However, you are not restricted to the last line. You may find it possible, in the course of a proof, to derive several lines within a single nec intro subproof to which you want to apply nec intro. The present statement of the rule permits you to do this without constructing a separate nec intro subproof for each. Let us illustrate this using the example of a conjunction that is necessary, where we want to derive the necessity of each conjunct.

1.	$\Box \, (p \, \& \, q \, \& \, r \, \& \, s)$	$/ \therefore \Box p \, \& \, \Box q \, \& \, \Box r \, \& \, \Box s$
2.	$\Box \,\vert\, (p \, \& \, q \, \& \, r \, \& \, s)$	T-reit
3.	$\quad p$	
4.	$\quad q$	
5.	$\quad r$	2, conj elim
6.	$\quad s$	
7.	$\Box p$	3 nec intro
8.	$\Box q$	4 nec intro
9.	$\Box r$	5 nec intro
10.	$\Box s$	6 nec intro
11.	$\Box p \, \& \, \Box q \, \& \, \Box r \, \& \, \Box s$	7–10 conj intro

(6) Note that where you have one nec intro subproof within the scope of another, you will not be able to T-reit a formula into the innermost subproof unless it has a □ remaining on it in the outer subproof. T-reit only allows you to cross *one* vertical line (□|) at a time. Each one you cross counts as a separate application of the rule, and so a □ must be removed each time.

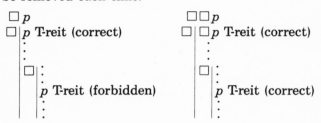

2.3.c   *Exercises*

A. Give proofs:

1. $p \rightarrow q$
    $q \rightarrow r$
    $\therefore p \rightarrow r$

2. $p \rightarrow q$
    $\Box p$
    $q \rightarrow r$
    $\therefore \Box r$

3. $\Box (p \,\&\, q)$
    $\therefore \Box q$

4. $\Box (p \lor q)$
    $\Box \sim p$
    $\therefore \Box q$

5. $\Box (p \lor q)$
    $p \rightarrow q$
    $\therefore \Box q$

6. $p \rightarrow q$
    $p \rightarrow \Box r$
    $q \rightarrow \Box \sim r$
    $\therefore \Box \sim p$

7. $\Box (p \lor (q \lor r))$
    $q \rightarrow s$
    $(p \lor r) \rightarrow t$
    $\therefore \Box (s \lor t)$

8. $(p \rightarrow r) \,\&\, (q \rightarrow r)$
    $\Box (p \lor q)$
    $\therefore \Box r$

(Note: The first premise is not of the form $\Box p$.)

We typically express the theorems of the various systems in terms of the lower case letters p, q, r, which are variables ranging over well-formed formulas of our symbolic language. Such expressions are schemata; any well-formed formula of the language exhibiting that logical structure is a theorem. For example, the expression $p \supset p$ is a theorem schema. The substitution of any wff for p yields a theorem:

$$A \supset A$$
$$((A \lor B) \equiv C) \supset ((A \lor B) \equiv C).$$

Keeping this in mind, we can see two things: (1) the theorems so written may be taken as generalizations about all sentences and may be read that way, e.g., '$p \supset p$' expresses the claim that every sentence implies itself; (2) a proof we give of a theorem schema is a proof schema for a proof of any substitution instance of the theorem schema.

B.   Prove the following theorems. Then try to state in English the general truth that each one expresses.

1. $p \rightarrow p$
2. $\Box\Box p \rightarrow \Box p$
3. $(p \rightarrow q) \rightarrow (\Box p \supset \Box q)$
4. $(\Box p \;\&\; \Box q) \leftrightarrow (p \leftrightarrow q)$
5. $\Box p \rightarrow (q \rightarrow p)$
6. $\Box \sim p \rightarrow (p \rightarrow q)$
7. $(p \rightarrow (q \;\&\; \sim q)) \rightarrow \Box \sim p$
8. $((p \rightarrow q) \;\&\; (\sim p \rightarrow q)) \rightarrow \Box q$

2.3.d   More Rules

It will be useful and convenient to have some rules that enable us to work with the possibility operator directly. Here we formulate introduction and elimination rules for possibility.

Possibility introduction

p
$\therefore \Diamond p$ poss intro

Possibility elimination

The name and formulation of this rule are due to Fitch. The rule is more intuitively viewed as a way of transmitting possibility than eliminating it. However we continue the pattern of designating the rules for our operators as "introduction" and "elimination" rules, even though it is less obviously appropriate here.

This rule enables us to introduce a possibility sign onto a sentence entailed by a sentence which is itself possible. The basis for the rule is the theorem

$$(p \rightarrow q) \supset (\Diamond p \supset \Diamond q),$$

which may be read as saying that whatever is logically entailed by something possible is itself possible.

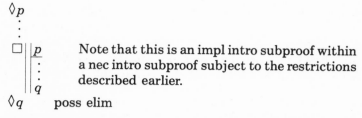

Note that this is an impl intro subproof within a nec intro subproof subject to the restrictions described earlier.

Greater fondness for this rule will result from trying to prove the theorem mentioned above and the theorems illustrated below without the rule. Here are some examples to illustrate proper use of the rule:

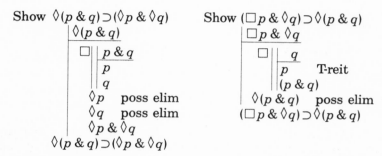

The rules just introduced are redundant in the sense they do not enable us to prove anything that we could not have proven without them. We can get the effect of these rules with the rules introduced prior to this section.

Your instructor may wish to add more rules of this sort to make it easier to give proofs and solve exercises. It is convenient to have some versions of modal modus ponens:

Modal modus ponens

$$p \rightarrow q \qquad\qquad p \rightarrow q$$
$$p \qquad\qquad\qquad \Box p$$
$$\therefore q \qquad\qquad\qquad \therefore \Box q$$

Exercises

1. Prove the redundancy of the rules just introduced in this section.
2. Show the redundancy of Rule R: If p is a theorem of propositional logic, then $\Box p$ is a theorem of propositional modal logic.

2.3.e   Decision Procedure

Our rules for T give us a way of demonstrating the validity of propositional arguments that involve modal concepts. We symbolize the argument using our connectives, capital letters to represent individual statements, and □ and ◊ to indicate the modality of the statements. Then we see whether or not we can derive the conclusion from the premises. If we can, the argument has a valid form. But if we can not, we do not know if this is because the form is invalid or because we were not sufficiently ingenious in trying to devise a proof.

In propositional logic, if we were unable to construct a proof, we could test the argument form by looking at the truth table for its corresponding conditional (or more quickly, by determining whether there is any consistent assignment of truth values that makes the antecedent true and the consequent false). The two-valued truth tables of propositional logic provide what logicians call an effective procedure for determining whether or not a given argument form is a tautology. That means that it is a mechanical procedure which will give an answer to your question in a finite number of steps. Now the truth tables were an effective procedure for telling that a formula is a theorem or a tautology, and they were effective for telling that a formula is not. In modal logic there is also a truth-table technique, but it is four-valued instead of two-valued, and though it is effective for telling that a given formula is not a theorem, the technique is not effective for telling whether a formula is a theorem of modal logic. This procedure is tedious and cumbersome, requiring 4^n rows in a table for a formula with n propositional variables.[7] So, for example, a formula with 3 different propositional letters would require a table with $4^3 = 64$ rows.

Saul Kripke, a leading contemporary writer on modal logic, has adapted a technique of E. W. Beth involving semantic tableaus. In Appendix III, we mention some authors who develop this semantic tableau technique for T, S4, and S5. While semantic tableaus are less tedious than quasi truth tables, they are better suited to establishing validity than uncovering invalidity.

2.3.f   Counterexamples (S5)

Here we will develop a technique for giving counterexamples. This technique, like that of devising proofs, depends on our ingenuity.

There is no series of steps for which we can give mechanical instructions and that will always produce a counterexample if the formula is invalid. So our failure to give a counterexample will not prove validity. But a counterexample is a decisive proof of invalidity. Although the technique has these limitations, it is interesting and useful from a philosopher's standpoint.

One limitation should be noted immediately. The technique to be described cannot discriminate T or S4 from S5. It can only be used to give counterexamples to formulas which are not theorems of any of these systems.

The intuitive idea for giving counterexamples is simple and easy to grasp. If someone claims that all A's are B's, and we point out an A that is not a B, then we have refuted the claim by giving a counterexample. Suppose, for example, someone claims that all the great philosophers were unmarried. We could give a counterexample by pointing out that Aristotle was married. Similarly, suppose someone claims that p necessarily implies q. We could give a counterexample to this by exhibiting a possible situation in which p is true while q is false. For example, if someone foolishly suggests that *my shirt is not red* entails that *my shirt is blue,* we can show that *my shirt is not red* does not entail that *my shirt is blue* in the way suggested. It is possible that I am wearing a green shirt. So it is possible that my shirt is not red while it is false that my shirt is blue. The basic idea here is simply to find an instance or type of instance that the universal generalization under consideration fails to cover.

This technique can be adapted to deal with modal formulas expressing inferences. Suppose we have before us the conditional formula corresponding to some alleged inference. Our efforts to demonstrate its validity have been unsuccessful, and we suspect it is invalid. We can prove its invalidity by displaying a false substitution instance – specific propositions which, when substituted properly into the argument form, yield a falsehood.

Example:

> The formula in question is $(\Diamond p \ \& \ \Diamond q) \rightarrow \Diamond(p \ \& \ q)$.
> Let $p = $ I am exactly six feet tall.
> Let $q = $ I am exactly five feet tall.

Each of these is possible and the conjunction of their *possibilities* is true, but the conjunction of p and q is clearly impossible. Hence the formula in this example is not a theorem.

It is also possible to give general counterexamples to modal formulas such as the one in the previous example. A general counter-example is one which uses propositional variables rather than specific propositions, and which in effect describes a large class of specific counterexamples. In giving general counterexamples we use these general assumptions:

a)   there is at least one contingent proposition $- \Diamond p \,\&\, \Diamond \sim p$;
b)   some contingent proposition is true $-p$;
c)   there is a pair of logically independent propositions $-$
     $\sim (p \rightarrow q) \,\&\, \sim (q \rightarrow p)$.

Our objective is to choose a general substitution instance of the inference pattern in question that

a)   makes the antecedent (or premises)
     $-$either a necessary truth (e.g., some obvious theorem)
     $-$or equivalent to an obvious assumption
     (note that this includes the conjunction of an assumption with a necessary truth);

and b)   makes the consequent (or conclusion)
     $-$either an obvious necessary falsehood
     $-$or equivalent to the denial of an assumption.

Example 1:
Consider the formula we treated earlier:

$$(\Diamond p \,\&\, \Diamond q) \rightarrow \Diamond (p \,\&\, q)$$

We have assumed that there is some proposition, p, that is contingent, i.e., $p \,\&\, \Diamond \sim p$. Consider a substitution instance in which p is substituted for p and $\sim p$ for q:

$$(\Diamond p \,\&\, \Diamond \sim p) \rightarrow \Diamond (p \,\&\, \sim p)$$

The antecedent is now equivalent to our first assumption, a), and the consequent is an obvious necessary falsehood. Consequently, we have a general counterexample to the alleged inference. Any contingent proposition and its denial will suffice.

Example 2: $(\Box p \,\&\, q) \supset \Box (p \,\&\, q)$

Let q be a contingently true proposition. This means we have q and $(\Diamond q \,\&\, \Diamond \sim q)$. Let p be the necessary truth $(q \lor \sim q)$. Substituting, we obtain

$$(\Box (q \lor \sim q) \,\&\, q) \supset \Box ((q \lor \sim q) \,\&\, q)$$

Looking at the consequent, we see that the expression within the scope of the necessity sign is a conjunction whose first conjunct is a tautology. Hence that expression is logically equivalent to the second conjunct alone. Making this substitution yields

$$(\Box(q \lor \sim q) \,\&\, q) \supset \Box q.$$

The antecedent is now true by virtue of the assumption that q is true and the fact that $q \lor \sim q$ is obviously necessary, and the consequent contradicts our choice of q as a contingent proposition, that is, $\Diamond \sim q$ or $\sim \Box q$.

Exactly why does this constitute a proof of invalidity? First of all, the theorems or candidates for theorems being examined are universal generalizations. Every permissible substitution instance of a formula must be true if it is a theorem. The form of every valid inference is a theorem. The technique just described provides a way of trying to discover a type of substitution instance of a formula that will show decisively that not all uniform substitutions within that formula are true, and therefore it is not a theorem of modal logic. Secondly, our substitution technique made certain assumptions. Is that proper here? These assumptions are consistent with using this as a disproof technique for modal logic because without these assumptions there is no point to having modal logic. Furthermore, these assumptions seem to be obvious truths.

Any propositional modal formula to which we can give a counterexample is not a theorem and does not correspond to a valid inference.

2.3.g   *Exercises*

A. Give proofs:
  1. $\Box p \rightarrow \Box (p \lor q)$
  2. $(p \rightarrow q) \supset (\sim q \rightarrow \sim p)$
  3. $\sim \Diamond p \supset \sim \Box p$
  4. $(p \rightarrow r) \,\&\, (q \rightarrow s)$
     $\Box p \,\&\, \Box q$
     $\therefore \Box r \lor \Box s$
  5. $(p \rightarrow r) \lor (q \rightarrow r)$
     $\Box (p \,\&\, q)$
     $\therefore \Box r$

B. Give proofs or counterexamples:
  1. $(p \rightarrow q) \supset (\sim \Diamond q \supset \sim \Diamond p)$
  2. $(p \rightarrow q) \supset (\Diamond \sim q \supset \Diamond \sim p)$

3. $(p \supset q) \supset (\sim \lozenge q \supset \sim \lozenge p)$
4. $(p \supset q) \supset (\lozenge \sim q \supset \lozenge \sim p)$
5. $[((p \ \& \ q) \rightarrow r) \ \& \ \square q] \rightarrow (p \rightarrow r)$
6. $[((p \ \& \ q) \rightarrow r) \ \& \ q] \rightarrow (p \rightarrow r)$
7. $\square p \rightarrow \square (p \ \& \ q)$
8. $\lozenge (p \lor q) \rightarrow (\lozenge p \lor \lozenge q)$
9. $(\lozenge p \lor \lozenge q) \rightarrow \lozenge (p \lor q)$
10. $(\sim \lozenge p \ \& \sim \lozenge q) \rightarrow \lozenge (p \lor q)$
11. $(p \rightarrow r) \lor (q \rightarrow r)$
    $\square (p \lor q)$
    $\therefore \square r$
12. $(p \rightarrow r) \ \& \ (q \rightarrow s)$
    $\square (p \lor q)$
    $\therefore \square (r \lor s)$
13. $(p \rightarrow r) \ \& \ (q \rightarrow s)$
    $\square p \lor \square q$
    $\therefore \ \square (r \lor s)$
14. $(p \rightarrow r) \lor (q \rightarrow s)$
    $\square p \lor \square q$
    $\therefore \ \square (r \lor s)$

C. Symbolize and prove:

1. If a proposition is possible, then its disjunction with any proposition is possible.
2. If each of two propositions is necessary, then their conjunction is necessary.
3. If a conjunction is necessary, then each of the conjuncts is necessary,
4. If either of two propositions is necessary, then their disjunction is necessary.
5. If one proposition is possible while another is not, then their disjunction is possible.

D. Symbolize and prove or give a counterexample:

1. If a disjunction is necessary, then either one or the other disjunct is necessary.
2. If a proposition is necessary, then its conjunction with any other proposition is possible.
3. If one proposition is necessary and another is possible, then their conjunction is possible.
4. If one proposition is possible and another is impossible, then their conjunction is impossible.

E. Prove or give a counterexample:
1. $(p \rightarrow (q \rightarrow r)) \rightarrow ((p \,\&\, q) \rightarrow r)$
2. $((p \,\&\, q) \rightarrow r) \rightarrow (p \rightarrow (q \rightarrow r))$
3. $((p \,\&\, q) \rightarrow r) \rightarrow (p \rightarrow (q \supset r))$
4. $(\Diamond p \,\&\, \sim\!\Diamond q) \supset \sim\!\Diamond (p \vee q)$ (compare C.5)
5. $((p \,\&\, q) \rightarrow r) \supset ((p \rightarrow r) \,\&\, (q \rightarrow r))$
6. $((p \,\&\, q) \rightarrow r) \supset ((p \rightarrow r) \vee (q \rightarrow r))$
7. $((p \vee q) \rightarrow r) \supset ((p \rightarrow r) \vee (q \rightarrow r))$
8. $p \rightarrow (q \rightarrow p)$
9. $p \supset (q \rightarrow p)$
10. $p \rightarrow (q \supset p)$
11. $(p \rightarrow (q \rightarrow r)) \rightarrow ((p \rightarrow q) \rightarrow (p \rightarrow r))$
12. $((p \rightarrow (q \rightarrow r)) \,\&\, (p \rightarrow q)) \rightarrow (p \rightarrow r)$

2.3.h *Iterated Modalities and Reduction of Modalities*

So far we have not dealt with formulas which have "piled up" or repeated modal operators, e.g.,

$$\Diamond \Box \Box \Diamond \Box p$$
$$\Diamond \Box \Diamond (\Box \Diamond p \supset \Box \Diamond \Diamond q)$$
$$\Diamond \Box (\Diamond \Box \Diamond p \,\&\, \Diamond \Diamond (\Box \Diamond \Box p \rightarrow q)).$$

Such formulas are difficult to understand. Our modal intuitions fail us when we try to make sense of them. What sense can we make, under our interpretation of the modal operators, of the difference between $\Box \Box \Box \Box p$ and $\Box \Box \Box p$? Thus it seems desirable to look for a way of reducing repeated modalities in such formulas, if possible.

First, we need to make more precise the concept of a repeated modality, or, as it is often called, an *iterated* modality. A sequence of modal operators containing two or more modal operators constitutes an *iterated* modality. There are several points to notice in connection with this concept. First, a well-formed formula can contain several iterated modalities, e.g., $\Box \Box \Box p \supset \Diamond \Diamond q$. Secondly, the occurrence of one modal operator within the scope of another does not necessarily constitute an iterated modality. For example, in $\Box (p \supset \Diamond q)$, although the $\Diamond$ occurs within the scope of the $\Box$, we do not have an iterated modality because it is not a sequence of modal operators containing two or more consecutive modal operators. Thirdly, when negations occur within a sequence of modal operators, as in $\Box \sim\! \Diamond \sim\! \Box p$, the sequence still is an iterated modality. This is because the negation always can either be removed or replaced by a single negation at the beginning of the sequence by substituting equivalences.

The example above is equivalent to $\square\square\square p$. The sequence $\lozenge \sim \lozenge p$ is equivalent to $\sim\square\lozenge p$. Finally, it is important to recall that our $\rightarrow$ abbreviates $\square(\dots \supset \dots)$. Recalling this enables us to see that $\lozenge(p \rightarrow q)$ contains an iterated modality $- \lozenge\square(p \supset q)$.

Any equivalence that enables us to reduce the number of modal operators in a sequence of iterated modalities by replacing it with a shorter sequence is called a *reduction law*.

These concepts enable us to put some of our questions more precisely:

1. How are we to understand (or read or interpret) iterated modalities?
2. Do we have any reduction laws in T?
3. Are there any intuitively correct reduction laws for the modal concepts as we are interpreting them that should be part of our system of modal logic?

Let us start with the first question. Consider a formula such as $\square\square p$ or $\lozenge\lozenge p$. Presumably these are to be read as

It is necessarily true that p is necessarily true,

and

It is possible that p is possibly true.

We recall that the sense of necessity involved is logical, not causal or epistemic. Now there does not seem to be any particular difficulty in understanding these formulas. They may seem peculiar, cumbersome, or even redundant, but they are not unintelligible. They may say things we would never have occasion to say in ordinary discourse or even philosophical discourse, but that is no drawback.

Formulas with more and more iterated modalities, however, are going to be more and more difficult to distinguish from each other by any means besides counting the modalities. That is, it is hard to conceive of what the difference in the number of modal signs between $\square\square\square\square p$ and $\square\square\square\square\square p$ is supposed to amount to. So while iterated modalities may not be unintelligible, it is difficult to see what the difference between certain ones is supposed to come to. Indeed, one is led to wonder whether there is any difference.

Are there any reduction laws in T? We know that we can prove $\square\square p \supset \square p$ and $\lozenge p \supset \lozenge\lozenge p$. If we could prove $\square p \supset \square\square p$ and $\lozenge\lozenge p \supset \lozenge p$, this would give us a reduction law that would enable us to reduce significantly many iterated modalities. But, alas, the latter formulas cannot be proven in T. So we do not have a logical equivalence, and therefore we lack a reduction law that would enable us to reduce consecutive □'s or ◊'s. In fact, there are no reduction laws in T. All modalities in T are irreducible.

Is that as it should be? Clearly the one theorem we have, $\Box p \supset p$, is acceptable. But are there other theorems that we should have (i.e., express modal truths) and would give us reduction laws but that we lack in T? Here we have before us a major issue distinguishing the systems T, S4, and S5. We have just observed that if we had

(AS4) $\Box p \supset \Box\Box p$,

we could then reduce any pair of □'s to a single □. Adding (AS4) as an axiom to T yields S4. AS4 is often called the "characteristic formula" of S4. Its presence in addition to what we have would give us the logical equivalence

$$\Box p \leftrightarrow \Box\Box p,$$

and it is a simple exercise to show that we also then have

$$\Diamond p \leftrightarrow \Diamond\Diamond p.$$

Is it true that if a proposition is necessarily true, then it is a necessary truth that it is necessarily true? If the necessity involved is logical necessity, then it is hard to think of any reasons for denying AS4. It seems very likely that any argument or example that establishes the contingency (or impossibility) of $\Box p$ will also establish the contingency (or impossibility) of p and thus fail to be a reason for rejecting the axiom AS4.

The explanation is quite clear in terms of the following "possible worlds" interpretation. Like Leibniz, let us say that a logically necessary truth is true in all possible worlds, and a logically possible proposition is true in at least one possible world. Is it possible for $\Box p$ to be true while $\Box\Box p$ is false, for some proposition p? This would mean that there is a possible world in which $\Box p$ is true and $\Box\Box p$ is false. If $\Box p$ is true in a world, then p is true in every possible world. But if $\Box\Box p$ is false in a possible world, then $\Box p$ is false in at least one possible world, and if there is a world in which $\Box p$ is false, then there is a world in which p is false. But now our supposition has led to the conclusion that there is a possible world in which p is true and p is false, which is impossible.

It appears, then, that our system T fails to include an important modal truth. We turn next to the system S4 which includes $\Box p \supset \Box\Box p$.

2.4   The System S4

The system S4 is one of eight modal systems devised by C. I. Lewis, one of the great modern pioneers of modal logic. It was not the system he favored, for reasons we will go into later. But it is something of a favorite among logicians, primarily because it is well-behaved and because it has some reduction laws.[8] Our natural deduction treatment of it is due to Frederick B. Fitch.

2.4.a   Rules

We have said earlier that S4 can be obtained from T by a simple modification of the rules. We have just now seen that this modification has to allow us to prove (AS4) $\Box p \supset \Box \Box p$, since the possibility of obtaining this formula distinguishes S4 from T. We can best see what modification is necessary by asking what prevents us from deducing AS4 in T. If we attempt a proof, the answer is soon obvious: It is the requirement that we must remove the $\Box$ when we reiterate a formula of the form $\Box p$ into a *nec intro subproof*. A good way to obtain S4, then, is introduce an S4-reiteration rule which drops that requirement.

Our rules of S4 will be all the rules of T plus the following S4-reiteration rule:

S4-reit: Where a formula of the form $\Box p$ occurs as a previous line in a proof, it may be entered as a line within a *nec intro subproof* with the justification 'S4-reit', unless that occurrence lies within the scope of a discharged assumption or within the scope of a terminated *nec intro subproof*.

This rule is to be used in conjunction with necessity introduction subproofs, just as the T-reit rule was. The T-reit rule is still available, but it becomes redundant with the addition of S4-reit.

We illustrate the use of this rule by proving AS4:

$$\begin{array}{ll} \Box p & \\ \quad \Box|\Box p & \text{S4-reit} \\ \quad \Box\Box p & \text{nec intro} \\ \Box p \supset \Box\Box p. & \end{array}$$

The way we modified the rules of T to obtain our rules for S4 makes it clear that T is *contained* in S4, in the sense that any theorem provable in T is provable in S4. And our earlier illustration with AS4 makes it clear that S4 is not contained in T.

2.4.b   Exercises

A. Show that T-reit is redundant in S4.

B. Prove:
   1. $\Diamond \Box \Diamond p \supset \Diamond p$
   2. $(p \rightarrow q) \supset (r \rightarrow (p \rightarrow q))$
   3. $\Box p \supset (q \rightarrow \Box p)$
   4. $(p \supset \Box q) \supset (p \supset (r \rightarrow \Box q))$
   5. $\Box \Diamond p \supset \Box \Diamond \Box \Diamond p$
   6. $\Box \Diamond \Box \Diamond p \supset \Box \Diamond p$
   7. $\Diamond \Box p \supset \Diamond \Box \Diamond \Box p$
   8. $\Diamond \Box \Diamond \Box p \supset \Diamond \Box p$

2.4.c   Reduction of Modalities in S4

In S4, we have the following equivalences, which are *reduction laws*:

RL1 $\Box p \leftrightarrow \Box \Box p$
RL2 $\Diamond p \leftrightarrow \Diamond \Diamond p$.

(RL2 is easily derived from RL1.)

From RL1 and RL2 we see that any time a modal operator has two consecutive occurrences in a formula, that formula can be replaced by an equivalent formula containing a single occurrence of the operator in that position. More interesting is the reduction of formulas containing mixtures of operators. These laws enable us to replace any formula containing iterated modal operators with an equivalent formula that contains no more than three iterated modal operators.

There are only two possible ways of combining three modal operators so that each operator stands next to a different operator:

$$\Box \Diamond \Box p$$
$$\Diamond \Box \Diamond p.$$

What formulas with fewer operators imply or are implied by these formulas? You have already provided a proof for one in exercise set B in section 2.4.b.

There are also two possible ways of combining four modal operators so that each operator stands next to a different operator:

$$\Box \Diamond \Box \Diamond p$$
$$\Diamond \Box \Diamond \Box p.$$

Each of these can be reduced to a formula in which p is prefixed by only two modal operators. Note that you have proven the appropriate equivalence for each of these in exercise set B at the end of section 2.4.b.

We see now that many iterated modalities can be reduced in S4. Four irreducible iterated modalities remain. But S5, the system we consider next, has no irreducible iterated modalities.

2.5   The System S5

S5, like S4, was first axiomatized by C. I. Lewis. S5 is a favorite system from a purely formal standpoint because it is a well-behaved system containing standard propositional logic and having no irreducible iterated modalities. But its primary virtue is being the formal system that best represents the notions of logical necessity, logical possibility, and logical entailment, where logical necessity is thought of as "*broadly* logical necessity" — the concept discussed at the beginning of the book. While the use of T and S4 would not commit us to any mistaken inferences involving these concepts, these systems do not provide us with the full range of inferences that are intuitively correct. We have the full range only in S5.

The characteristic formula of S5 is

$$\Diamond p \supset \Box \Diamond p.$$

In S5 modal status is always necessary, and the characteristic formula of S5, together with the characteristic formula of S4 (which is deducible in S5), can be seen as saying this. The characteristic formula of S4, $\Box p \supset \Box \Box p$, says that if a proposition is necessary, its necessity (i.e., its modal status as a necessary proposition) is necessary. Likewise, the characteristic formula of S5 may be seen as the counterpart claim for possible propositions. $\Diamond p \supset \Box \Diamond p$ says that if p has the modal status of being a possible proposition, it is necessary that it is a possible proposition.

S5 best reflects and expresses the possible worlds account of necessary truth, the view that a necessary truth is one that is true in all possible worlds. This interpretation confirms the plausibility of the characteristic formulas of S4 and S5. Consider the characteristic formula of S4 first. Suppose we have some proposition, p, that is true in every possible world. Since each possible world contains every proposition true in that world, each world contains not only p but also the proposition that p is true in every world. If every world contains the proposition that p is true in every world, it is true in every world that p is true in every world, i.e., it is necessary that p is necessary.

If we read the characteristic formula of S5 in this way we have: if p is true in some world, then in every world p is true in some world. If we are talking about broadly logical necessity it is very hard to see how some proposition could be true in some world (i.e., possible), and yet fail to be a logical possibility in some other world (i.e., not necessarily possible). If it is possible that there are unicorns, then there is some possible world in which it is true that there are unicorns. But then it is true in every world that it is true in some world that there are unicorns. That is to say, it is necessary that it is possible that there are unicorns. So the characteristic formula of S5 expresses something intuitively acceptable, something there seems to be no good reason to deny.

2.5.a   Rules for S5

The only difference between the rules of T and the rules for S4 comes in the restriction on the reiteration rule. The T-reiteration rule allows only lines of the form $\Box p$ to be reiterated into nec intro subproofs and specifies that when these lines are reiterated, the $\Box$ must be dropped. The S4-reiteration rule likewise allows only lines of the form $\Box p$ to be reiterated, but it does not require that the $\Box$ be dropped. We can obtain a rule for S5 by relaxing the restrictions still more.

A look at the characteristic formula of S5 will show us how to relax the restriction further so as to obtain the proper rule. The characteristic formula of S5

$$\Diamond p \supset \Box \Diamond p$$

implies that modal status is always necessary. If a proposition is necessary, then it is necessary that it is necessary, and if a proposition is possible, it is necessary that it is possible. The formula $\Diamond p \supset \Box \Diamond p$ contains the clue to the modification needed in our rules to get S5. Clearly what we must do is relax the restriction that only formulas of the form $\Box p$ may be reiterated into *nec intro* subproofs so as to permit reiteration of formulas of the form $\Diamond p$ as well.

We obtain the system S5 by adopting an S5-reit rule which relaxes this restriction. Our rule is stated as follows:

S5-reit: Where a modal formula (a formula of the form $\Box p$ or of the form $\Diamond p$) occurs as a previous line in a proof, it may be entered as a line within a nec intro subproof with the justification 'S5-reit,' unless that occurrence lies within the scope of a discharged assumption or within the scope of a terminated nec intro subproof.

The statement of the reiteration rule for S5 makes it clear that T and S4 are contained in S5 and that S5 is not contained in either of them. T is the weakest of our systems; S4 is stronger than T but weaker than S5.

2.5.b   Exercises

A. Prove the following formulas using the weakest reiteration rule that still permits derivation of the formula.
   1. $\Box(p \vee \Box q) \supset (\Box p \vee \Box q)$
   2. $\Box(p \vee q) \supset (\Box p \vee \Diamond q)$
   3. $\Box(p \vee q) \supset \Box(\Box p \vee \Diamond q)$
   4. $\Box(p \,\&\, \Box q) \supset (\Diamond p \,\&\, \Diamond q)$
   5. $(\Box p \vee \Box q) \supset \Box(p \vee \Box q)$
   6. $\Diamond(p \,\&\, \Box q) \supset (\Diamond p \,\&\, \Box q)$
   7. $(\Diamond p \,\&\, \Diamond q) \supset \Diamond(p \,\&\, \Diamond q)$
   8. $(\Diamond p \,\&\, \Box q) \supset \Diamond(p \,\&\, \Box q)$
   9. $\Box(p \vee \Diamond q) \supset (\Box p \vee \Diamond q)$
   10. $(\Box p \vee \Diamond q) \supset \Box(p \vee \Diamond q)$

B. Prove that by adding $\Diamond p \supset \Box \Diamond p$ as an axiom to our system T, the characteristic formula of S4, $\Box p \supset \Box \Box p$, follows.
   Hints: 1) Any axiom is itself necessary, i.e., $\Box(\Diamond p \supset \Box \Diamond p)$;
   2) any wff with the general form $\Diamond p \supset \Box \Diamond p$ is an axiom; (Any formula that has *that* structure, where p is any wff you like, is an axiom.)
   3) Be sure to consider the axiom in both of its guises, i.e., both $\Diamond \Box p \supset \Box p$ and the above form.

2.5.c   Reduction Laws

We saw earlier that we could reduce the number of consecutive modal operators in many formulas, given the rules of S4. We did this with the help of the reduction laws based on AS4:

RL1 $\Box p \leftrightarrow \Box \Box p$
RL2 $\Diamond p \leftrightarrow \Diamond \Diamond p$.

Although we could significantly reduce the number of iterated operators where long strings of them occurred, there were certain irreducible combinations in S4. In S5 we can always reduce a sequence of iterated operators to a single operator.

We can carry on such a reduction by continuing to reduce pairs of iterated operators, if we have a way of reducing any possible pair. Now there are only four possible pairs,

$$\Box\ \Box$$
$$\Diamond\ \Diamond$$
$$\Box\ \Diamond$$
$$\Diamond\ \Box.$$

A glance back at RL1 and RL2 shows that we have appropriate reduction laws for the first two on the list. Furthermore, we have a start on proving the required equivalences in the case of the latter two. We have

$$\Box\Diamond p \supset \Diamond p$$
$$\Box p \supset \Diamond\Box p$$

in T and S4. Now in S5, we are able to obtain equivalences and gain the remaining reduction laws. Indeed, the formula needed to prove the necessary equivalence is the characteristic formula of S5. So we may add to our stock of reduction laws in S5.

RL3 $\Box p \leftrightarrow \Diamond\Box p$
RL4 $\Diamond p \leftrightarrow \Box\Diamond p$

All the iterated combinations that were irreducible in S4 are now reducible to a single modal operator in S5. In fact, whenever we have iterated modal operators in S5, we may remove all the modal operators except the last one.

There is also another type of reduction that can be carried out in S5. To describe this, we introduce the concept of modal *degree* of a formula. With iterated modalities, this concept works just as one would expect: $\Box\Diamond\Box p$ is a formula of degree 3; $\Box\Diamond\Box\Diamond p \supset \Box\Diamond p$ is a formula of degree 4. This concept also applies to formulas in which a formula containing modalities itself occurs within the scope of modalities, even though no iteration occurs, e.g., $\Box(\Box p \supset \Box p)$ has degree 2. The concept of degree encompasses iterated modalities and other situations as well. The degree of a formula can be precisely calculated as follows:

1)   a propositional variable has degree 0;
2)   if p has degree n, then $\sim p$ has degree n;

3)   if p has degree n and q has degree m, $p \vee q$, $p \,\&\, q$, and $p \supset q$ have the degree of the larger component;

4)   if p has degree n, then $\Box p$ and $\Diamond p$ have degree $n + 1$.[9]

Not only can all iterated modalities in S5 be reduced to a single modal operator, but every formula of a degree greater than one can be reduced to a first-degree formula in S5.

2.5.d Exercises

A. Prove:

1. $\Box \Diamond \Box p \equiv \Box p$
2. $\Diamond \Box \Diamond p \equiv \Diamond p$
3. $\Box \Diamond \Box \Diamond (p \rightarrow \Box \Diamond q) \supset (\Box \sim p \vee \Diamond q)$
4. $\sim (r \rightarrow \Diamond \sim p) \supset (\Box p \,\&\, \Diamond r)$
5. $(\sim (r \rightarrow \Diamond \sim p) \,\&\, (\Diamond r \rightarrow q)) \supset ((\Box p \vee \Diamond \sim q) \,\&\, \Diamond r)$

B. Give the degree and reduce to first degree formulas:

1. $\Box \Diamond \Box \Diamond (\Diamond \Diamond p \supset \Box \Diamond p)$
2. $\Box (p \vee \Box q)$
3. $\Diamond (\Box p \vee \Diamond \Diamond q)$
4. $\Diamond (p \,\&\, \Box q)$
5. $\Diamond (\Box \Diamond p \supset q)$

2.6   Philosophical Matters

2.6.a Differences between Systems

So far we have emphasized the formal aspects of the differences between the Systems T, S4, and S5. They differ in size — that is, S5 has more theorems than S4, and S4 has more than T. T fits inside S4, and S4 is contained in S5. S5 is therefore called the strongest of these three systems, and T is the weakest of the three. These three systems also differ in the reduction laws that are available in each system. T has none, S4 has some, while S5 permits reduction of the modality of any formula to a first degree modality.

But these are largely formal differences. They do not give a philosophical "picture" of how the systems differ. Now we turn to a more intuitive picture for thinking of the differences. The way of looking at the differences presented here is due to Saul Kripke and has become a popular way of dealing with this matter. It uses the idea of possible worlds and of an *accessibility* relationship among them. The different modal systems will represent differing degrees of accessibility among possible worlds. But that is getting ahead of the story. First we build our picture and introduce the necessary terminology.

A possible world is a maximally consistent state of affairs, a total way a universe could be. We imagine an array of such possible worlds. We will call a proposition *true in a possible world* just in case it would be true if that world were actual. It will be convenient to think of possible worlds mainly in terms of the propositions true in them.

Setting up our array of possible worlds, we further suppose that we are given some relationship R among these worlds. Some may be related by R while others are not. We will think of this relationship as "accessibility" or "relative possibility." Depending upon how it behaves, we will have pictures or models of one system or another. We call a proposition *possible in a world* just in case it is true in some world accessible to the world in question. Similarly, a proposition is *necessary in a world* just in case it is true in every world accessible to that given world. Depending on how various worlds are accessible to others in a given model, we might have a proposition necessary in one possible world but not in another, or a proposition that is possible in one world but not in another. The theorems of modal logic are generalizations about relationships of propositions within one array of possible worlds. By altering properties of the accessibility relation we can give models of T that do not make true all the theorems of S4, and models of S4 that do not make true all the theorems of S5, as well as models satisfying S5. The way this works depends only on structural features of the model – upon properties of the accessibility relation among possible worlds – and not upon which individual propositions turn out to be true or false in a given possible world. The theorems of T, for example, turn out to be true in all models having a certain structure, regardless of the truth or falsity of individual propositions in those models. What kinds of structures correspond to each of our systems?

We start with T. Its characteristic formula is $\Box p \supset p$. First we give a model in which this formula *is true,* and then we give further pictures in which this formula is true but the characteristic formulas of S4 and S5 are not. The minimal network to model $\Box p \supset p$ requires exactly one possible world accessible to itself in which $\Box p$ is true. Given some possible world accessible to itself with $\Box p$ true in it, we see that p holds also, because $\Box p$ means that p is true in every accessible world. While it may seem trivial to state that the accessibil-

ity relation is reflexive here, that is precisely the feature of accessibility guaranteeing that $\Box p \supset p$ will be true.

But this picture is not very exciting as an attempt to model T,
because $p \supset \Box p$ is also true in it. This means that our model so far
fails to distinguish T from propositional logic. Given both $\Box p \supset p$ and
$p \supset \Box p$, we would have the reduction law, $\Box p \equiv p$, that would allow
us to eliminate all modalities, and thus our modal logic would collapse back into classical propositional logic. We want a more "interesting" model, one that not only distinguishes T from classical propositional logic but from S4 and S5 as well.

Let us suppose that we have a network of three possible worlds
accessible or inaccessible to each other as indicated in Figure 1.

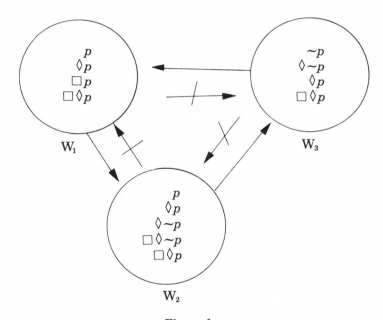

Figure 1

→ = is accessible to, where a world W_1 is accessible to a world W_2 just in
case any proposition true in W_2 is possible in W_1.

Arrows represent accessibility and slashed arrows indicate lack of
accessibility. We assume that each world is accessible to itself, although this is not indicated by arrows on our diagram. Inside each

circle is a list of some propositions true in the possible world represented by that circle. Let us see what happens to some of our characteristic formulas in this model.

The characteristic formula of T, $\Box p \supset p$, holds here. Wherever $\Box p$ is true, so is p. This can be seen directly in W_1. Two other *instances* of this characteristic formula, $\Box \Diamond \sim p \supset \Diamond \sim p$ and $\Box \Diamond p \supset \Diamond p$, also hold in W_2 and W_3, respectively, and there are no counterexamples. Indeed, given that any proposition that is necessary in a world is true in every world accessible to that world, and that every world is accessible to itself, it is easy to see that there cannot be any counterexamples.

This model enables us to distinguish T from propositional logic, something which the first model failed to do. For p holds in W_2 while $\Box p$ does not, since there is an accessible world, W_3, in which p is not true. Thus we have a counterexample (actually a countermodel) to the formula that would be needed to "collapse" modality, $p \supset \Box p$.

The characteristic formulas of S4 and S5 also fail in this model. $\Box p$ occurs only in W_1, and $\Box \Box p$ does not hold there because $\Box p$ does not appear in W_2, the only other world accessible to W_1. Thus $\Box p \supset \Box \Box p$, the characteristic formula of S4, does not hold in this model.

The characteristic formula of S5, $\Diamond p \supset \Box \Diamond p$, appears at first glance to be satisfied, since $\Diamond p$ appears in each world and so does $\Box \Diamond p$. However, in W_3 we find that $\Diamond \sim p$ is true while $\Box \Diamond \sim p$ is not. $\Box \Diamond \sim p$ is not found in W_3 because $\Diamond \sim p$ does not occur in W_1, the other world accessible to W_3. Thus the formula $\Diamond \sim p \supset \Box \Diamond \sim p$ is not true in the model, and since it is an instance of $\Diamond p \supset \Box \Diamond p$, the characteristic formula of S5, we have here a model which fails to satisfy S5, even though initially it might have appeared to satisfy it.

The accessibility relation in Figure 1 is reflexive; that is, every world is accessible to itself. But this relation is not symmetric there; W_1 is accessible to W_2, but W_2 is not accessible to W_1. Nor is it transitive: W_1 is accessible to W_2, and W_2 is accessible to W_3, but W_1 is not accessible to W_3.

What would happen if the accessibility relation were transitive as well as reflexive? As the reader may suspect, we will have S4 models. Or, to put it differently, we will not be able to devise models in which the accessibility relation is reflexive and transitive and in which the characteristic formula of S4 fails to hold. To illustrate this, we will use a model similar to the one in Figure 1, except that we alter the accessibility relations somewhat. We want a model of T and S4 in which the characteristic formula of S5 fails to hold.

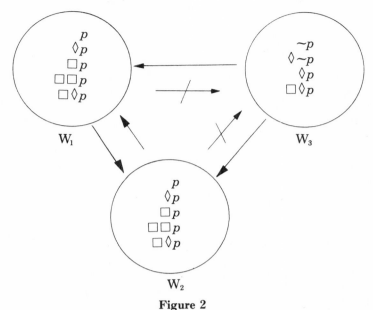

Figure 2

[→ = is accessible to]

First we check the characteristic formulas of T and S4. In each
world in which we have a formula of the form $\Box p$, we have a formula
of the form p. The same is true for the characteristic formula of S4.
We have formulas of the form $\Box\Box p$ in W_1 and W_2, and there we have
$\Box p$ in each case. Now what of $\Diamond p \supset \Box \Diamond p$? Again, although initially
it seems to be satisfied, it is seen to fail. Once again $\Box\Diamond p$ appears
in each world where $\Diamond p$ appears, seeming to satisfy $\Diamond p \supset \Box\Diamond p$. But
W_3 contains $\Diamond\sim p$ without containing $\Box\Diamond\sim p$, which means that not
all instances of $\Diamond p \supset \Box\Diamond p$ hold in this model. Where the accessibility relation is reflexive and transitive but not symmetric, the characteristic formula of S5 is not guaranteed to hold.

If the accessibility relation is symmetrical as well as reflexive and
transitive, then it becomes an equivalence relation on the set of possible worlds, and every possible world is accessible to every other.
In fact, as Kripke remarks, S5 may be viewed as a modal system that
requires that all the possible worlds be accessible to each other.

This way of looking at the differences between the systems in terms
of the properties of the accessibility relation suggests the possibility
of another system "between" T and S5. Suppose that the accessibil-

ity relation is reflexive and symmetric, but not transitive. In such a case we would have a system containing T but not S4, and contained in S5, and whose characteristic formula is $p \supset \Box \Diamond p$. In fact, adding this formula as an axiom to the axioms of S4 yields S5 — exactly what we would expect given the relationships we have just considered. This formula is sometimes called the *Brouwersche* axiom, although the connection with Brouwer is tenuous.[10]

Comparing these various modal systems in terms of different properties of the accessibility relationship offers a way of seeing why S5 is the modal system that best represents the understanding of broadly logical necessity. That which is broadly logically necessary is true in all possible worlds, true no matter what. When we use this picture of possible worlds and accessibility relationships, the necessity of a proposition in a given world — in the actual world, for example — is the truth of that proposition in all possible worlds accessible to the actual world. But this would not coincide with truth in all possible worlds unless the accessibility relation were reflexive, symmetric, and transitive. So if we think of broadly logically necessary truths as true in *all* possible worlds, S5 seems to be the most adequate modal system.

So far, all we have done is argue that modal inferences properly follow certain patterns. We have not tried to say which propositions are broadly logically necessary, nor have we tried to devise any test for picking them out. Many controversial candidates for necessary propositions have been proposed by philosophers, but to accept S5 (or any other modal system) does not commit us to accept any philosopher's candidates for broadly logically necessary truths, beyond the theorems of S5 (or of the other modal system accepted).

2.6.b Some Assumptions Implicit in Our Presentation

When we accept S5 as the most adequate and accurate account of the logic of broadly logical possibility and necessity, we are making a number of assumptions about propositions and modality. We have seen in our discussion of accessibility that adopting S5 rules is presuming that broadly logical necessity is a kind of unlimited necessity — for something to be necessary in this sense is for it to be true in every possible world, *every* possible state of affairs. We do not admit as a possibility that some proposition should be necessary in one possible world and not be necessary in another. We assume that whatever is broadly logically necessary in the actual world is necessary in every possible world. But, as we have just observed, this does not imply anything about our ability to *know* whether or not a given proposition is necessary, nor does it imply that we can provide a completely reliable test by which we can discover whether a proposition is necessary.

In order to reveal some further assumptions we have been making, let us consider a criticism of all the systems we have introduced (and more besides) made by A. N. Prior. This criticism applies to all systems that contain as a theorem the formula

$$\Diamond p \supset \Diamond (p \vee q)$$

i.e., if it is possible that p, then it is possible that p or q. Prior tries to show that this is false by giving counterexamples. You might wonder how Prior could possibly given an example in which it is true that p is possible and false that p or q is possible. But this is not exactly what he does.

> ... let p be 'Only God exists,' and suppose this possible, and let q be 'I don't exist.' Here p is possible *ex hypothesi*, but could the disjunction 'Either p or q,' i.e.,
>
> 'Either only God exists or I don't exist,'
>
> possibly be true? The peculiarity of it is that the disjunction is unstatable unless I exist, and is therefore only statable if both parts of it, 'I don't exist' and 'Only God exists,' are false. In this case, then, 'Possibly either p or q' is false although 'Possibly p' is true.[11]

Prior's contention, notice, is that the proposition in question is statable (by anyone other than God) only if it is false, so it is not possible for it both to be stated and true. Well, you might say, what does it matter whether or not anyone can *state* the proposition truly? After all there *is* such a proposition, and if there is, then it *is* possibly true even if I cannot state it. But this reply assumes just what Prior wishes to reject—that there is such a proposition whether statable or not. Perhaps he is harking back to the medieval idea of regarding a proposition as a *dictum,* a thing said. Thus he assumes that where something is unstatable, we do not have a *dictum* or proposition, and therefore since propositions are the bearers of truth or falsity, we do not have truth or falsity. It is worth noting here that Prior holds this not only for cases of pronomial reference, such as "I don't exist," but also for *dicta* using names which fail to refer—that is, names of nonexistent objects. Thus "Santa Claus doesn't exist" likewise expresses a *dictum* that is either false or unstatable, and similarly lacks a truth value where it is unstatable. This *dictum* cannot actually be asserted in such a way that the referring expression actually refers and the assertion is true.

Looking back at the inference Prior is questioning, we see he holds that it is possible for the antecedent to be true and possible for the consequent to fail to be true. The consequent fails to be true, not because it is false, but because it is not statable and hence lacks truth value.

Prior's alleged counterexample exposes an important assumption. Accepting the counterexample entails accepting the claim that at least some propositions are contingent beings – i.e., they can fail to exist. We have tacitly assumed that all propositions are necessary beings – that is to say, they all exist and have truth values in all possible worlds, although of course not all are necessary truths. Now we can see the importance of this assumption to the adequacy of the modal systems we have been considering.

Although we will not defend our assumption here, we may explore a bit further some consequences of denying it. The most important consequence is that we must revise our accounts of possibility and necessity, since there are additional cases to be dealt with.

We described necessary propositions as those true in all possible worlds, and we can continue to use that characterization. But, if we continue to hold that necessity and possibility are interdefinable while we allow the possibility that propositions fail to exist in some worlds, the definition of possibility must be changed. Let us call those propositions true in all possible worlds *strongly necessary*. Then the corresponding sense of possibility becomes *weak possibility* – the not-strongly-necessarily false. This means that any propositions that are not false in every possible world are weakly possible. These include not only the strongly necessary, and those true in some worlds but false in others, but also those propositions that are true in every world in which they exist but do not exist in all worlds, as well as those false in every world in which they exist but that do not exist in all worlds. Equivalently, we could say that a proposition is weakly possible just in case there is some world in which the denial of that proposition fails to be true.

But now there is a second way to understand these modal notions of possibility and necessity. Essentially it involves changing where we put the cases in which the proposition fails to exist in some worlds. Let us call a proposition *weakly necessary* if and only if there is no world in which that proposition exists and is false. This, of course, counts a proposition as necessary even though it is not true in every possible world. Correlatively, a proposition is *strongly possible* if and only if there is some possible world in which it is true. (If there are no propositions that exist in some worlds and not in others, then these two accounts of necessity and their corresponding senses of possibility are simply equivalent.)

This gives us two distinct pairs of modal notions:

strong necessity – weak possibility
weak necessity – strong possibility.

A proposition would be weakly necessary but not strongly necessary if there were some worlds in which it failed to exist while it was true in every world in which it did exist. Prior apparently thinks the propositions we express when each of us says, "I exist," are of this sort. Likewise, if a proposition that fails to exist in some worlds is false in every world in which it exists, then it is weakly necessarily false. Presumably "I do not exist" expresses a proposition of this sort for Prior.

Having distinguished these pairs of modal concepts, we can see that Prior's alleged counterexample works only if possibility is taken as *strong possibility*. That is, the modal operators in the formula $\Diamond p \supset \Diamond (p \vee q)$ must be interpreted in the sense we have called strong possibility in order for the theorem to turn out false. The formula continues to be true for weak possibility. So for Prior's counterexample to work, it must be possible for some propositions to fail to exist, and possibility must be defined as strong possibility.

But there is a strange result if we follow Prior here. In the passages where this counterexample is presented,[12] Prior suggests that in worlds where I don't exist, the propositions I now express by saying "I exist" and "I don't exist" also fail to exist. If that were the case, then the proposition I express by assertively uttering the words "I exist" is a (weakly) necessary truth. This seems to be an unhappily implausible consequence if here we are trying to capture the notion of "broadly logical necessity" described earlier.

On the other hand, defining possibility as weak possibility and adopting Prior's position on "I exist" yields the conclusion that contradictions are (weakly) possible. Since I exist now, the propositions I express by saying "I exist" and "It's false that I exist" both exist. Hence their conjunction, a proposition normally taken to be necessarily false or impossible, exists. But since I presumably fail to exist in some other worlds, this conjunction does not exist in some other possible worlds. However, it then follows by the definition of *weakly possible* that this conjunctive proposition is possible. Hence a proposition that exists and is usually taken to be impossible is (weakly) possible, given Prior's assumptions.

There are probably many lines to pursue on this matter. Although

we will not pursue them here, we will be forced to look at similar issues later on in dealing with modality and quantification when we
deal with the existence of individuals. Now we turn to another question raised not only by these conundrums but also by the fact that
there are so many systems of modal logic: Is there one correct system of modal logic?

2.6.c   Is There Only One Correct System of Modal Logic?

We have before us three systems of modal logic, and there are many
more systems which we have not considered. Are they all correct?
Is only one correct?

In an Aristotelian Society symposium addressed to these questions, E. J. Lemmon[13] suggests an analogy with a situation in mathematics. There are many systems of geometry: Euclidean geometry,
and a host of non-Euclidean geometries. Now there is one weak sense
in which all of these geometries are correct: They are all *consistent*
formal systems. They can each be modelled so that all the theorems
of the system turn out to be true in the model. But which one is true
of actual physical space? Which is true of the lines and triangles and
circles, etc., with which we have to do in our daily lives? It seems that
only one system can be correct and that it is an empirical matter to
find out which one.

Alas, the question is not quite so straightforward as this. For to
answer the question in this way, we must make some assumptions
about the accuracy of our measurements – usually that our instru-
ments remain constant at different times and places. But, continues
Lemmon,

> If we make this assumption, we shall find, according to Einstein that
> in fact there are triangles whose angles do not add up to two right angles:
> on this assumption, physical space does not conform to Euclidean ge-
> ometry. On the other hand, we can preserve the Euclidean model for
> physical space by sacrificing the assumption, by admitting instead that
> one's measuring instruments are deformed in certain ways at different
> points.[14]

Thus, which geometry turns out to be correct depends on other theoretical considerations. Any answer to the question requires a full
description of the situation, and even if we have as full a description as we can obtain, the facts at our disposal may not dictate the
answer.

The analogy contains several morals applicable to deciding which modal logic is correct. First, it is not sufficient to ask about the formal systems alone or even the formal system interpreted with the help of an abstract model, such as a set theoretical model. We must ask about the system under its intended interpretation, some actual application. Then we can sensibly ask whether one system rather than another is correct. Secondly, our intended interpretation must be clear enough to enable us to answer questions about the truth or falsity of theorems once they are interpreted. And, finally, we must consider the effect adopting the system will have on other theories we hold (or want to hold).

There is another consideration important to our present case. We have before us three systems that are all compatible. The larger ones contain the smaller ones. So if S5 is correct, both S4 and T contain no falsehoods; rather, they do not contain all the modal truths. In such a situation T and S4 would not be incorrect but insufficient. Yet the correct system should be as adequate as possible; it should contain all the relevant truths.

Although only one system is most adequate under an intended interpretation, and we have suggested that S5 is the preferred system where necessity is understood as broadly logical necessity, it does not mean that there is no interest in the other modal systems, including ones not presented here. For there are other ways of understanding necessity, as we have just seen in our brief consideration of Prior, and there are still other quite different interpretations of modal operators that philosophers have considered. Relative to these other interpretations, one of the other modal systems may be the correct one. Let us look at some of the alternative interpretations to get an idea of the possibilities.

We have mentioned earlier that C. I. Lewis, the "inventor" of S4 and S5, rejects these systems in favor of a weaker one, S2. Lewis began exploring modal logic because of his dissatisfaction with '⊃' as the formalization of the concept of implication, as used in *Principia Mathematica* and in nearly all symbolic logic textbooks today. He thinks that one proposition implies another when it is impossible both for the implying proposition to be true and for the implied proposition to be false. He further thinks of implication in terms of *deducibility*. Lewis uses a fishhook as his symbol for implication and he interprets the expression $p \multimap q$ as meaning that q is deducible from p. However he *defines* $p \multimap q$ in the same way we have defined $p \rightarrow q$ —as $\Box(p \supset q)$.

He employs his interpretation when he rejects S4 and S5 for containing $(p \multimap q) \multimap ((q \multimap r) \multimap (p \multimap r))$ as a theorem:

> I doubt whether this proposition should be regarded as a valid prin-
> ciple of deduction: it would never lead to any inference $p \dashv 3\, r$ which
> would be questionable when $p \dashv 3\, q$ and $q \dashv 3\, r$ are given premises; but
> it gives the inference $(q \dashv 3\, r) \dashv 3\, (p \dashv 3\, r)$ whenever $p \dashv 3\, q$ is a premise.
> Except as an elliptical statement for "$((p \dashv 3\, q)\ \&\ (q \dashv 3\, r)) \dashv 3\, (p \dashv 3\, r)$ and
> $p \dashv 3\, q$ is true," this inference seems dubious.[15]

Lewis is here claiming that this formula is at least dubious and perhaps false when interpreted as being about deducibility. He does not
doubt that when q is deducible from p and r is also deducible from
q, it can be deduced that r is deducible from p. But he is reluctant
to believe that when q can be deduced from p, we can always go on
to deduce that r's deducibility from p is deducible from r's deducibility from q.

In S4 and S5

$$\Box[\Box(p \supset q) \supset \Box(\Box(q \supset r) \supset \Box(p \supset r))]$$

is provable as a theorem, and it seems acceptable under our interpretation. This suggests that Lewis either rejects our interpretation
of $\Box$ and $\Diamond$, or that he rejects other principles that are needed for
the deduction of this formula. I think it is most plausible to suppose
that he is reading $\Box$ as something other than broadly logical necessity, since he formally defines $p \dashv 3\, q$ the very same way we define
$p \rightarrow q$. He obviously takes the concept of necessity to be more closely
connected with deducibility than we do.

Arthur Prior has studied tense-logical interpretations of several
modal systems.[16] One way of obtaining a tense-logical interpretations
of the modal systems we have considered is to read the operators as
follows:

> $\Diamond p =$ it is or was or will be the case that p
> $\Box p =$ it is, was, and will be the case that p (or it is always
> the case that p)
> $p =$ it is (now) the case the p.

Obviously some of the basic axioms are true on this interpretation.

> $\Box p \supset p =$ If it is, was, and will be the case that p, then it
> is (now) the case that p;
> $\Box(p \supset q) \supset (\Box p \supset \Box q) =$ If it is always the case the $p \supset q$,
> then if it is always the case that p, then it is always the
> case that q.

So is R – if p is a theorem of propositional calculus, then p is, was, and
will be the case. Other tense-logical interpretations are also possible.

Jaakko Hintikka has developed an epistemic interpretation using the structure of S4.[17] Here the operators are read

$\Box p =$ Smith (or some specific individual) knows that p (written $K_s p$);

$\Diamond p =$ It is possible for all Smith knows that p (written $P_s p$);

while $p =$ it is true that p.

Obviously it is important to keep careful track of the subscripts on the "modal" operators, since most of the inferential relationships will hold only within one given individual's knowledge and belief, and not between the beliefs of different individuals. For example, *Smith's* knowing p will imply that *he*, Smith, believes p, but Smith's knowing p will imply nothing about Jones's beliefs.

a) $\Box p \equiv \sim \Diamond \sim p$ is interpreted as
$K_s p \equiv \sim P_s \sim p$, which is read as "Smith knows p iff it is not possible, for all S knows, that $\sim p$";

b) $\Box p \supset p$ is interpreted as
$K_s p \supset p$, which is read as "If Smith knows p, then p is true";

c) $\Box (p \supset q) \supset (\Box p \supset \Box q)$ is interpreted as
$K_s (p \supset q) \supset (K_s p \supset K_s q)$, which is read as "If Smith knows that p implies q, then if Smith knows that p, then Smith knows that q";

d) $\Box p \supset \Box \Box p$ is interpreted as
$K_s p \supset K_s K_s p$, which is read as "If Smith knows p, then Smith knows that he knows p."

The wary reader may be suspicious of some of these, but we will not examine them further here. Suffice it to say that Hintikka is aware of the possible grounds for suspicion and he defends his position on these points. Needless to add, he develops this interpretation with more skill and subtlety than is exhibited here.

We could continue examining interpretations of modal systems,[18] but the point is clear. Many interpretations can be given for the various systems. When we ask whether one system is correct or which system is correct, we must always ask those questions relative to some interpretation. Different interpretations are required for different purposes, and different systems are likely to be preferred for different interpretations.

3

QUANTIFICATION

Before we turn to quantified modal logic, we must specify which symbols and rules we will use for predicate logic. The system of predicate logic – quantifier logic, quantification theory, or first-order logic – used here is the same as that presented in nearly any standard introductory textbook. Despite differences in symbols and in the ways rules are specified, exactly the same formulas turn out to be theorems and the same argument forms turn out to be valid. Here are the symbols and rules used in this text.

3.1 Symbols

Quantification theory adds two operations to those employed and symbolized in propositional logic.

Operation	Symbol	Example	Other commonly used symbols
Universal Quantification	(x)	$(x)Ax$	$(\forall x)$, Vx, Πx
Existential Quantification	$(\exists x)$	$(\exists x)Ax$	(Ex), Λx, Σx

Lower case letters from the end of the alphabet are used as variables, and Roman capitals with a variable or variables written immediately to the right represent what are variously called predicates, sentential functions, open sentences, propositional functions and relations. For example, the expression Ax might represent the predicate *x is an aardvark*, and Axy might symbolize *x is the aunt of y*.

We also adopt the usual characterization of free and bound occurrences of variables. The occurrence of a given variable will be *bound* in a given formula when it either occurs in a quantifier or within the scope of a quantifier containing that variable. An occurrence of a variable that is not bound is a free occurrence.

3.2   Interpretation

The universal quantifier is designed to capture certain ordinary uses of "all" and "every," and the existential quantifier represents that use of "some" which is more precisely expressed as "at least one." Where $Gx = x$ is a Greek and $Hx = x$ is human, $(x)(Gx \supset Hx)$ symbolizes the statement that all Greeks are humans and $(\exists x)(Gx \,\&\, Hx)$ symbolizes the statement that some Greeks are human (or, more precisely, that at least one Greek is human).

We will understand the quantifiers to range over existing objects. It may seem strange that we bother to say that, but there are two matters at stake here.

First, there are two ways to read quantifiers – the "objectual" way just mentioned, and a substitutional way. The substitutional way construes formulas of the form $(\exists x)\,(ax)$ as follows: There is a true substitution instance of the formula (open sentence) ax. This has the consequence that uses of existential quantifiers carry with them no implications about what objects there are. The objectual view of quantification, on the other hand, reads such formulas so as to imply that there exists an object of which the predicate is true when we have a true existential statement. The difference between these views can be clarified with an example. It is true that Fafner (or Santa Claus) does not exist. According to the substitutionalist such a truth licenses the existential generalization, $(\exists x)\,(x \text{ does not exist})$. Or similarly, from Bucephalus no longer exists, we may infer $(\exists x)\,(x$ no longer exists). The objectualist demurs in these and similar cases. By objectualist lights, $(\exists x)\,(x \text{ does not exist})$ expresses something self-contradictory and so any inference of this from a truth must be invalid.[19]

The second matter at stake here is related to the first and will come up again later. It is the question of what we quantify over – the *range* of the quantifiers. We can restrict the range of our quantifiers. For example, in an arithmetic textbook, we might understand our quantifiers to be ranging only over integers, or over rational numbers, or over real numbers. In such a case the formula $(\exists x)\,(x$ is prime or x is over 7 feet tall) is understood to mean that there is a *number* that is either prime or over seven feet tall. It is clear that *in such a context* the open sentence, x is prime or x is over 7 feet tall, is satisfied by the number three and not by Ralph Sampson. It is possible, of course, to expand our universe of discourse beyond numbers. We may add sets, persons, material objects – anything that we would call a concrete object or an abstract object. Is it possible to expand to our universe of discourse beyond the abstract and concrete objects which actually exist? Well, there have been proposals that we quantify over fictional entities or over possible but nonactual objects. Here, however, we will be conservative and limit ourselves to existing things. This proposal is not intended to settle any philosophical arguments about what does and does not exist; we will consider that to be beyond the scope of our concern here.

3.3   Equivalences

Like the modal operators, the quantifiers are interdefinable. We will officially consider the universal quantifier to be undefined and the existential quantifier to be defined in terms of it. (We could, if we chose, do it the other way around.) Here is a list of equivalences, which may be used in proofs, expressing the relationships between the quantifiers.

$$(\exists x)\, \alpha x \equiv\, \sim (x) \sim \alpha x$$
$$(x)\, \alpha x \equiv\, \sim (\exists x) \sim \alpha x$$
$$\sim (\exists x)\, \alpha x \equiv (x) \sim \alpha x$$
$$\sim (x)\, \alpha x \equiv (\exists x) \sim \alpha x$$

We will label our use of these rules QE (for Quantifier Equivalence).

3.4   Rules

In quantification theory we have an expanded means of symbolizing and two additional operations (just one if we count only primitives). To continue our natural deduction treatment, we require introduction and elimination rules for the universal quantifier and for the existential quantifier.

Statements of rules vary from text to text. Certain restrictions are necessary, but it is possible to introduce them in different ways, and most textbooks employ some orthographic device to insure compliance with the restrictions. The intuitive idea behind each rule usually is simpler than the technical statement of the rule.

Let us use a special symbol to facilitate the statement of our rules. αx will stand for a formula possibly containing free occurrences[20] of the variable x. $(x)\, \alpha x$ will be the universal quantification of αx.

Universal Instantiation (UI)

From $(x)\,\alpha x$
infer αy (or αx)   where y is free and replaces all free occurrences of x in αx.

Universal Generalization (UG)

From αy (1) where αx is obtained from αy by re-
infer $(x)\,\alpha x$ placing all free occurrences of y by x;
             (2) where no free occurrences of x in αy are bound in $(x)\alpha x$ (This restriction is to prevent us from accidentally binding a variable that should remain free.);
             (3) where y does not occur free in an assumption within whose scope αy (the formula being generalized) lies. (This includes EI assumptions; see below.)

This set of restrictions is a technical (syntactic) way of preventing certain fallacious inferences. Just remember these rules of thumb: Do not apply UG to a variable introduced by EI; and do not apply UG to a variable introduced in an assumption until you have discharged the assumption.

Existential Generalization (EG)

From αx where we need not replace all occur-
infer $(\exists y)\alpha y$ rences of x by y.

Existential Instantiation (EI)

From $(\exists x)\,\alpha x$
use $\left|\begin{array}{l}\alpha y \text{ EI asp}\\ \vdots\\ p\end{array}\right.$ (1) where αy is the result of replacing all free occurrences of x in αx with y;
             (2) where y is a variable that does not occur in any line of the proof prior to its
to infer p EI concl   introduction in the EI subproof, and y does not occur free in p.

This rule appears more complicated than the others. However, this form of the rule with its restrictions fits some common sense insights about the reasoning that involves Existential Instantiation. The first is that if it is true that something has property a, then it is permissible to assume that some arbitrarily selected individual in the universe of discourse has a and draw conclusions from that assumption, provided that we use no special features of the individual chosen. Secondly, our assumption is just that — an assumption. We do not *infer* that this arbitrarily selected individual has a, and the conclusion we draw in closing the EI proof must not involve that "instance." Our form of the EI rule explicitly represents these insights.

The "EI asp" line captures the first insight. We *assume* that some individual has a. The restriction on the choice of the variable insures that the choice is arbitrary and prevents the use of any feature peculiar to the individual chosen.

The second insight is captured by the use of a subproof that permits deductions from our chosen "instance," but that does not permit any conclusion containing that "instance" to be entered as a line outside the subproof. Thus the rule permits us to assume an instance of an existential generalization while it forbids us to bring conclusions about that particular instance outside the scope of the assumption.

The statement given of the EI rule can be regarded as an abbreviation of a use of implication introduction and UG together with a familiar theorem about the scopes of quantifiers. Consider the following proof schema:

$(\exists x)a\,x$

$\quad\;\; a y$	(considered this time as
$\quad\;\; \cdot$	an assumption for impl
$\quad\;\; \cdot$	intro)
$\quad\;\; p$	
$a y \supset p$	Impl intro
$(x)(a x \supset p)$	UG
$(x)\,(a x \supset p) \supset [(\exists x)a x \supset p]$	(Theorem)
$(\exists x)a x \supset p$	MP
p	MP

The restrictions on the EI rule are just those needed to guarantee that this proof will always work. If the subproof were not carried out in terms of a "new" variable, then it would not be guaranteed that UG could be used. And if p contained any free occurrences of the variable introduced in the subproof, then the theorem about quantifiers could not be used.

This shows that the form of EI adopted here is both formally and intuitively a good one for a natural deduction system.

For those who need to become accustomed to these rules, we present some sample proofs and offer a few exercises. An important "trick" to keep in mind when using the EI rule is to deduce within the EI subproof the conclusion you are working for and then bring it outside of the subproof.

Prove $(x)(Ax \supset Bx) \supset ((\exists x) \sim Bx \supset (\exists x) \sim Ax)$

$(x)(Ax \supset Bx)$	Asp
$(\exists x) \sim Bx)$	Asp
$\sim Bx$	EI asp
$Ax \supset Bx$	UI
$\sim Ax$	MT
$(\exists x) \sim Ax$	EG
$(\exists x) \sim Ax$	EI concl
$(\exists x) \sim Bx \supset (\exists x) \sim Ax$	impl intro
$(x)(Ax \supset Bx) \supset ((\exists x) \sim Bx \supset (\exists x) \sim Ax)$	impl intro

Prove $(x)(Fx \supset Q) \supset ((\exists x)Fx \supset Q)$
(Q here represents any formula containing no free occurrences of x.)

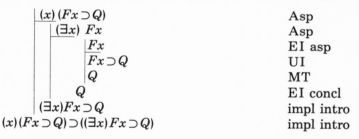

$(x)(Fx \supset Q)$	Asp
$(\exists x) Fx$	Asp
Fx	EI asp
$Fx \supset Q$	UI
Q	MT
Q	EI concl
$(\exists x)Fx \supset Q$	impl intro
$(x)(Fx \supset Q) \supset ((\exists x)Fx \supset Q)$	impl intro

Prove $((\exists x)Fx \supset Q) \supset (x)(Fx \supset Q)$

$((\exists x)Fx \supset Q)$	Asp
Fx	Asp
$(\exists x)Fx$	EG
Q	MP
$Fx \supset Q$	impl intro
$(x)(Fx \supset Q)$	UG
$((\exists x)Fx \supset Q) \supset (x)(Fx \supset Q)$	impl intro

Exercises

1. $(p \supset (x)Qx) \supset (x)(p \supset Qx)$
2. $(\exists x)(p \supset Qx) \equiv (p \supset (\exists x)Qx)$
3. $(x)(Ax \supset Bx) \supset [(x)(\sim Ax \supset \sim Cx) \supset (x)(\sim Bx \supset \sim Cx)]$

Since the UI rule permits the deduction of formulas containing free variables, it is necessary to add a final word about the theorems of quantifier logic. The theorems are any *closed* formulas deducible by means of the rules given here, including the earlier rules for propositional logic. A closed formula is one that contains no free occurrences of any variable. Of course it is a simple matter to close a formula we deduce containing free occurrences of some variables: we apply UG, or if we like, EG.

4

QUANTIFIED MODAL LOGIC

This chapter presents quantified modal logic within the framework of propositional modal logic and quantification theory already given. After a brief look at the need for quantified modal logic, we examine some traditional modal distinctions that have an important bearing on our development of quantified modal logic. The most important is the distinction between modality *de re* and modality *de dicto*. We use this distinction to explain what our modal signs, □ and ◊, mean when applied to predicates, and we discuss translation of modal claims from natural language into our symbolic language. After a brief look at some objections to quantified modal logic, we discuss and summarize the philosophical considerations we are trying to satisfy in the system presented here.

With these philosophical considerations in hand we consider three versions of quantified modal logic that have a legitimate claim to the title "Quantified S5." The first and easiest to develop is a version studied by Saul Kripke in 1959. Both it and a subsequent modification by Kripke in 1963 have unhappy philosophical consequences, however. A third version, called Actualism, is expounded and defended as the best system of quantified modal logic relative to a possible worlds interpretation. We present a natural deduction version of actualism equivalent to an axiomatization by Thomas L. Jager. This system is discussed and used by Alvin Plantinga in *The Nature of Necessity* and other writings.

The last three sections of the chapter show how the system of quantified modal logic developed here provides superior resolution of certain problems associated with modal logic.

4.1 Motivation

Why extend modal logic by adding quantification? What benefits do we expect? Are there inferences we can shed light on that are not illuminated by the methods developed so far?

We develop our answer with the help of an example.

1.   All men are necessarily rational.
2.   Socrates is a man.
3.   ∴ Socrates is necessarily rational.
4.   ∴ Socrates is (actually) rational.
5.   ∴ Socrates is possibly rational.

The necessity has to be represented formally in (3) in order to account for the inference of (4) and (5). Yet we cannot say that it is a necessary truth that Socrates is rational (since it is *not* a necessary truth because Socrates could have failed to exist). On the other hand (3) follows from (1) and (2), which are both true, so it is true; and it implies (4) and (5).

This example suggests two conclusions. First of all, the necessity here indicated figures in the inference. We cannot simply say that this argument uses a predicate, "necessarily rational," that needs no special representation. While this suggestion might help us explain how (3) follows from (1) and (2), we cannot account logically for the inference of (4) and (5). The other conclusion suggested by the example is that the sense of necessity here, while perhaps related to broadly logical necessity, is not the same.

Some explanation of necessity, in this present sense, needs to be given and it should be related to the sense introduced for propositional modal logic. This will be done here by considering a pair of old distinctions common to medieval discussions of modality. Then it will be easier to pick out the appropriate sense of necessity and relate it to broadly logical necessity.

4.2   Two Distinctions

The venerable distinctions are (a) the *de re/de dicto* distinction and (b) the distinction between the *composite* sense and the *divided* sense of modal terms. These distinctions will be considered only as they apply to our modal terms, possibility and necessity. It will be clear that each distinction has a far wider range of applicability.

Medieval discussions of modal arguments frequently appeal to a distinction between modality *de re* and modality *de dicto*. Modality *de re* is modality thought of as applying to a thing (*res*), more precisely, as a way a thing possesses a property. For example, one thing might be said to possess a property necessarily, or something can be said to possibly have some given property, as in the claims that Socrates is necessarily rational or Socrates is possibly a sailor.

Modality *de dicto* is the modality applied to a statement (*dictum*). It refers to the manner or mode of a statement's being true. For example, it is necessarily true that all bachelors are unmarried. Here it is the statement (the *dictum*) that is said to be necessary. More exactly, it is the statement's being true that is necessary.

Often an ordinary sentence, such as "Socrates is rational necessarily," is ambiguous. The word "necessarily" might be taken to refer to Socrates's possession of rationality, as in

Socrates is necessarily rational,

or to the necessity of the statement, as in

It is necessary that Socrates is rational.

Since the former is true and the latter is false, it is obvious that there is some point to keeping them distinct, and that the result of confusing these two senses in an argument can be disastrous.

The *de dicto* sense of necessity can be readily identified with broadly logical necessity, as this was described earlier. That is not to say, however, that broadly logical necessity is what the Medievals who discussed *de dicto* modality always had in mind. Further consideration of *de dicto* modality will be postponed until after we look at a related matter.

I turn first to another modal distinction, one which is occasionally linked with the *de re/de dicto* distinction and sometimes confused with it. This is the distinction between a modality taken in the composite sense (*in sensu composito*) and one taken in the divided sense (*in sensu diviso*). The *locus classicus* for explanations of this distinction is a passage in Aristotle's *De Sophisticis Elenchis* (166a):

> For the meaning is not the same if one divides the words and if one combines them in saying that 'it is possible to walk-while-sitting' [and write while not writing]. The same applies to the latter phrase, too, if one combines the words 'to write-while-not-writing': for then it means that he has the power to write and not to write at once; whereas if one does not combine them, it means that when he is not writing he has the power to write.[21]

Resorting to our symbolism, we can represent Aristotle as distinguishing

$$\Diamond(\exists x)\ (x \text{ is a man } \&\ x \text{ is not writing } \&\ x \text{ is writing})$$

(taking the phrase in composition or compositely) from

$$(\exists x)\ (x \text{ is a man } \&\ x \text{ is not writing } \&\ \Diamond(x \text{ is writing}))$$

(taking the phrase in division or dividedly and assuming prior to the explanation that we can make some sense of this use of the ◊). The importance of making this distinction is clear when we observe that the latter statement is true while the former is necessarily false.

It should be observed here that this distinction is not the same as the *de re/de dicto* distinction, but is simply a distinction between two ways of representing the scope of a modality. It may appear that because the first symbolization of Aristotle's example has the modality applying to a *dictum* and the second to a thing's possession of a property, the distinctions come to the same thing. But we can illustrate the distinction between the composite and the divided sense, eliminate any *de dicto* reading, and probably more accurately represent Aristotle's meaning as follows:

$(\exists x)$ (x is a man & $\Diamond$(x is writing & x is not writing))
(composite sense)

as compared with

$(\exists x)$ (x is a man & x is writing & $\Diamond$(x is not writing))
(divided sense).

Similar illustrations can be provided where the modality is *de dicto* throughout. For example,

It is possible for it not to be raining when it is raining

can be read

compositely as $\Diamond(\sim R \,\&\, R)$
and dividedly as $(\Diamond \sim R) \,\&\, R$.

Thus it is clear that the distinction between taking a modality in the composite sense and the divided sense is different from the *de dicto/ de re* distinction, although they may coincide in many cases.

These distinctions are linked with the previously discussed distinction between the necessity of the consequent and the necessity of the consequence by St. Thomas Aquinas. In a well-known passage in his *Summa Contra Gentiles,* he invokes these distinctions to resolve the question of whether God's foreknowledge of a contingent action (or truth) necessitates that action (or truth):

> If each thing is known by God as seen by Him in the present, what is known by God will then have to be. Thus, it is necessary that Socrates be seated from the fact that he is seen seated. But this is not absolutely necessary or, as some say, with the *necessity of the consequent*; it is necessary conditionally, or with the *necessity of the con-*

sequence. For this is a necessary conditional proposition: *if he is seen sitting, he is sitting.* Hence, although the conditional proposition may be changed to a categorical one, to read *what is seen sitting must necessarily be sitting,* it is clear that the proposition is true if understood of what is said *de dicto,* and compositely; but it is false if understood of what is meant *de re,* and dividedly. Thus in these and all similar arguments used by those who oppose God's knowledge of contingents, the fallacy of composition and division takes place.[22]

Thomas's resolution of the problem here seems to be correct, but the apparent conflating of distinctions is mistaken.

Thomas here points out that the claim

> If God sees that Socrates is seated, then necessarily, Socrates is seated

is ambiguous. It could mean

> Necessarily, if God sees that Socrates is seated, then Socrates is seated,

which expresses the necessity of the consequence, or it could mean

> If God sees that Socrates is seated, then it is a necessary truth that Socrates is seated,

which expresses the necessity of the consequent. Although the first reading expresses something true, it does not yield the conclusion that it is a necessary truth that Socrates is seated, given the additional premise that God sees that Socrates is seated. On the other hand, the latter reading, from which the undesirable conclusion does validly follow, is false. Thus Aquinas uses the distinction between necessity of the consequence and necessity of the consequent to show that the proffered argument is invalid or unsound. Either way we are spared the conclusion he wants to avoid.

The examples just considered make it clear that we have here three distinct pairs of concepts, which may overlap in their applications. St. Thomas, in the passage just considered, seems to think that in his example the distinctions coincide, that applying any one of them gives the same result. But it is not clear whether Aquinas thinks this is true in general. The distinctions between the composite and divided senses and between *de re* and *de dicto* are apparently more at home when applied to categorical propositions. So Aquinas turns the ambiguous hypothetical into an equivalent ambiguous categorical proposition to make his point. Thus

> What is seen sitting must necessarily be sitting

is true if understood compositely, i.e., with necessity applying to the composite:

Necessarily, what is seen sitting is sitting.

Obviously this is also a *de dicto* reading. The modality is external to the statement; it is applied to the dictum. However, understanding Aquinas's ambiguous example in the divided sense means applying the modality to the predicate without simultaneously applying it to the subject—

What is seen sitting is sitting necessarily.

Here the thing (*res*) that is seen to be sitting, whatever it might be, is said to have the property of sitting and to have it in a necessary manner or mode. In the present example, this claim is false, since what is seen to be sitting is Socrates and he does not have the property of sitting necessarily.

Despite the coincidence of the three distinctions in St. Thomas's example, it remains clear that we cannot claim that in general these distinctions will coincide or that each distinction has the same basis. The distinctions between necessity of the consequent and necessity of the consequence, and that between the composite and the divided sense, seem simply to be distinctions of scope of modalities. It is not true in general that the terms of these pairs pick out a different kind of modality—that "necessarily" or "possibly" have different meanings, depending on whether the modality is understood compositely or dividedly.

Returning now to our distinction between modality *de re* and modality *de dicto*, we might think that here we have a distinction between two different kinds of modality—between the manner in which some individual has a property and the manner in which a statement (dictum) is true. Statements of modality *de re* frequently make use of the adverb 'essentially' in place of 'necessarily.' This association with essences has, in part anyway, served to make the notion of modality *de re* suspect. Some philosophers who are suspicious of essences have suggested that modality *de re* is intelligible or acceptable only if it can be explained in terms of modality *de dicto*. On the other hand, such eminent medieval logicians as Peter Abelard, Peter of Spain, and William of Sherwood thought of modality *de re* as basic, choosing to explain *de dicto* modality in terms of the essential possession of truth by a statement.

Here I will make no attempt to decide which type is basic, if indeed either one should be considered "reducible" to the other. Likewise, I will not attempt to refute arguments against the legitimacy of either kind of modality. And I will not try to assess the basis of either kind of modality. All of these issues are interesting topics, worthy of further investigation.[23]

We are going to take *de re* modality to express the essential possession of a property by an individual. We use the □ to represent *de re* as well as *de dicto* modality. In our symbolism, we will continue to regard an occurrence of a modality sign as representing modality *de dicto* if and only if that formula and only that formula within the scope of modality contains no free occurrences of any individual variables. The easiest way to apply this test is as follows: Delete all parts of the whole formula except for what appears within the scope of the modal sign under consideration. If what remains contains no free occurrences of any individual variable, the modality is *de dicto*. Correspondingly, a modality is considered *de re* if and only if the formula within the scope of the modality does contain at least one free occurrence of an individual variable. Hence, in the following example, the first (outermost) occurrence of □ is modality *de dicto* and the second occurrence of □ is modality *de re*:

$$\Box(x) \ \Box \ (Cx \supset Hx).$$

In giving rules for quantified modal logic, we frequently will be forced to take stands on complex and hotly disputed metaphysical issues involved in understanding essence and existence, and *de re* and *de dicto* modality. Part of our purpose is to exhibit how and where these issues arise, and what the consequences are of various positions that can be taken on these issues.

Before we can see what rules it would be appropriate to have for □ in the context of quantification theory, we must first make sense of applying the □ to predicates. Prior to this chapter we have been reading (interpreting) □ as meaning "necessarily" or "it is necessarily true that." But this reading will not do when we apply the □ to predicates (or to symbolic representations of predicates). Reading $(x) \Box \, \alpha x$ as "for anything x, x *has* α is a necessary truth," is unsatisfactory. For one thing, the expression 'x has α' does not express a truth (or a falsehood), let alone a necessary one. So how can an operator like □, as we have been reading it, be sensibly attached to it? The answer seems to be that it cannot. On the other hand, if we read the □ as "essentially," and say, "for any x, x has α essentially," then we are using a different interpretation of the □, and we must justify our continued use of the symbol □ and establish a clear and compatible connection with our earlier *de dicto* interpretation.

Our strategy here is the latter approach. This strategy amounts to assuming that the logic of "essentially" and of essential properties, parallels that of "necessarily" and of necessary truths, and that we can find some straightforward logical connections between them. Much of what is done subsequently is designed to justify or at least vindicate these assumptions.

4.3 Symbolization

How to translate philosophical idioms into symbols becomes increasingly subtle and difficult in quantified modal logic. These problems frequently center on the scope of the modality and problem of rendering claims about properties in first-order logic. Consideration of an example will illustrate the point.

In Meditation VI, Descartes announces that the body is essentially divisible while the soul is essentially indivisible. Consider the first claim, that the body is essentially divisible. How is this claim best understood?

1) $\Box(x)$ (x is a body $\supset x$ divisible)[24]
2) (x) (x is a body $\supset \Box(x$ is divisible))
3) $\Box(x)$ (x is a body $\supset \Box(x$ is divisible))
4) $(x)\Box(x$ is a body $\supset x$ is divisible)
5) $(x)\Box(x$ is a body $\supset \Box(x$ is divisible))

(1) is a straightforward *de dicto* reading. It is said to be a necessary truth that everything that is a body is divisible, or, alternatively, that there is a necessary (essential) connection between being a body and being divisible. (2) says something quite different: that everything that is a body, (regardless of whether it is a body accidentally or essentially) has the property of being divisible essentially, although there need not be any necessary connection between having a body and being essentially divisible. To put it a bit differently, (2) might be true even if it is *possible* for something to be a body without being essentially divisible, as long as everything that *is* a body is essentially divisible.

The third reading (3) combines the first two. It says that it is a necessary truth that anything that is a body is essentially divisible. It excludes the possibility just mentioned above.

The fourth reading is best compared with the first. It is a *de re* reading, unlike (1), and it says that given any thing there is, it is an essential property of it that it is divisible if it is a body. The essential property here is a hypothetical property – an "if-then" property. For any thing there is, it could not have had a body and yet been indivisible. But to say that everything there is is essentially such that if it is a body then it is divisible does not preclude the possibility that there might have been something that was a body but nevertheless was indivisible. Saying that it is a necessary truth that whatever is a body is divisible does preclude this latter possibility. Reading (1) implies (4), but (4) does not imply (1).

In reading (5), just like (4), we have an essential connection between two properties with the additional feature that one of the properties is said to be essential. (5) says that for all the things there are (but not necessarily for all the things there *could* be), none of them *could* both be a body and fail to be divisible essentially. For any thing there is, being essentially divisible is essentially connected with (implied by) its being a body.

Which of these readings might Descartes have had in mind when he says that the body is essentially divisible? Obviously we would have to take account of the context in the *Meditations,* Descartes's metaphysics, and even of the history of metaphysics to produce a definitive answer to that question. I happen to think that the first reading is the best, but I believe a case could be made for some others as well.

It is worth noting here while we are thinking about what these readings mean, that (3) is the strongest. It implies each of the others, and no one of these others implies it. Once we have a set of rules in place, we will be able to confirm this intuition in our symbolic system.

Exercise: Symbolize, giving what you regard as the most plausible reading. Be able to explain your symbolization. Some of the exercises are genuinely ambiguous. Give the various possible readings in such cases.

(1)   Everything exists necessarily.
(2)   Everything exists essentially.
(3)   Possibly persons have bodies, but they can exist without them.

4.4   Some Objections to Quantified Modal Logic

Of those philosophers who have objected to quantified modal logic, one of the most notable is W. V. O. Quine. Quine's central objection is that if we permit quantification into *de dicto* modal contexts, we generate fallacies, but if we interpret those contexts as expressing *de re* modality, then we are committed to "Aristotelian Essentialism," which for him is an unacceptable metaphysical view.

Quine does not couch his objections in terms of a *de re/de dicto* distinction. He prefers to speak of modality or necessity that attaches to a thing as contrasted with modality or necessity that attaches to a way of speaking. In Quine's view, "... necessity resides in the way we say things, and not in the things we talk about."[25] Necessity that resides in the way we say things is, for Quine, a barely acceptable way to interpret *de dicto* necessity, analytic truth, and Lewis's strict sense of necessity. Quine observes that

> According to the strict sense of 'necessarily' and 'possibly,' these statements would be regarded as true:
>
> (15) 9 is necessarily greater than 7,
> (16) Necessarily if there is life on the Evening Star then there is life on the Evening Star,
> (17) The number of planets is possibly less than 7,
>
> and these as false:
>
> (18) The number of planets is necessarily greater than 7,
> (19) Necessarily if there is life on the Evening Star then there is life on the Morning Star,
> (20) 9 is possibly less than 7.[26]

But, he continues, there is a serious problem with these contexts of modal terms,

> ... for substitution on the basis of the true identities:
>
> (24) The number of planets = 9,
> (25) The Evening Star = the Morning Star
>
> turns the truths (15)–(17) into the falsehoods (18)–(20).[27]

In making his objection, Quine takes what we have called the *de dicto* reading of modal proposition as the only reading that is philosophically acceptable. He then observes that substitution of identities into *de dicto* modal contexts enables us to infer falsehoods from truths by otherwise impeccable rules of inference, and he concludes that we ought not to permit quantification into *de dicto* modal contexts.

But the advocate of quantified modal logic need not disagree with Quine on the points just made. Indeed advocates of quantified modal logic typically do not want to quantify into *de dicto* modal contexts. Rather they disagree with Quine about the propriety of recognizing *de re* modalities — modalities attaching to things. This is a disagreement, as Quine observes, about the tenability of "essentialism," the views that things have some properties essentially and others contingently. Quine flatly rejects this position, although it is difficult to see what his reasons are beyond the already stated claim that modality attaches only to statements, not to things.

Those who find essentialism true or even tenable can distinguish two claims in

> Nine is necessarily greater than seven.

One is the *de dicto* claim that

> It is a necessary truth that nine is greater than seven;

the other is the *de re* claim that

> Nine has the property of being greater than seven essentially or necessarily.

It is this latter claim that is thought to provide a context into which we may legitimately quantify and substitute.

In short, Quine's objections to quantified modal logic do not center on purely formal aspects of quantified modal logic, but on a certain interpretation of the modal operations. Since we have already said we do not intend to interpret the □ as indicating *de dicto* modality when the □ has as its scope an open sentence, we need not worry about Quine's objection that quantification is illegitimate and that substitutivity fails in those contexts. Instead, we must turn to the task of explaining more clearly what we mean by calling these contexts *de re* and reading the □ as "essentially."[28]

Before we turn directly to the question of essential properties, we should remember the situation. When we dealt with propositional modal logic, we considered the meaning of the modal operator before we turned to formal systems, and we judged formal systems by their adequacy to the primary meanings of the modal terms. At this stage of the discussion we already have first-order logic and we already have a system of modal logic, S5 (or S4 or T, for those who prefer the weaker systems). We will pursue the "finer" questions of interpretation and of metaphysics by combining the modal systems with first-order logic and raising our questions in a formal and philosophical context, rather than by trying to anticipate and answer questions prior to introducing the formalism. This procedure may also have an advantage in helping us identify the questions on which we will have to take stands, and enable us to avoid (or not, at our pleasure) some controversies that do not have to be resolved in order for us to agree on quantified modal logic.

4.5   Essential Properties and Possible Worlds

We have been thinking about modality in terms of possible worlds. But we must not think of a possible world as some kind of concrete object sitting out there waiting to be inspected. Nor should we think of it as some shadowy domain of possible but non-actual objects. Possible worlds are abstract entities: maximally consistent sets of statements. Given that statements exist and exist as abstract entities independent of human thought, possible worlds exist all right, and exist independently of whether or not we choose to think about them. Their existence, however, is the same sort of existence we ascribe to sets or to numbers.

We have described necessary statements as being true in all possible worlds and possible statements as those true in at least one possible world. That description was fine for the *de dicto* modalities considered in chapter II. But now our project is to make sense of *de re* modal claims such as

(1)   Socrates is necessarily (or essentially) rational

and (2)   Socrates possibly has a Roman nose,

or general statements like

(3)   All men are essentially rational

and (4)   Some men are possibly snubnosed.

This involves saying something about what it means for an individual to exist in a possible world and to have properties in a possible world.

Perhaps the most straightforward initial suggestions for understanding statements like (1) and (2) are to take (1) as telling us that Socrates is rational in every possible world and (2) as telling us that Socrates is Roman-nosed in at least one possible world. Furthermore, these suggestions seem to have the additional advantage of allowing us to reduce *de re* modal statements to *de dicto* ones. For example, the statement that Socrates is necessarily or essentially rational could be read as saying that the statement that Socrates is rational is true in every possible world. Likewise, to say that Socrates is possibly Roman-nosed could be taken as equivalent to saying that in some world it is true that Socrates has a Roman nose.

If we choose to adopt this course, we must face up to certain consequences. One of the first is that contingent entities cannot be said to have essential properties. Presumably rationality is one of Socrates' essential properties. However, it is not true in every possible world that Socrates is rational, and hence, according to the reading we are considering, Socrates is not essentially rational. Pursuing the point just a bit further, it seems that the reason why there are worlds in which it is not true that Socrates is rational is that Socrates does not exist in those worlds. He is a contingent being, existing in some worlds and not others. Obviously it will be true in general that contingent beings cannot have essential properties, given this present *de dicto* way of understanding statements about essential and possible properties. This result, however, runs counter to the strong intuitions many philosophers have had about essential properties.

Some philosophers have suggested that individuals can have properties in worlds in which they do not exist. On this suggestion, then, it might be true in every world that Socrates is rational even though in some worlds it is false that Socrates exists. Again, however, this is a suggestion that runs counter to the intuitions of many philosophers.

The best course of action at this point, it seems, is to look for an alternative way of understanding having a property essentially. But before we do that let us make sure we understand some of these locutions that we have begun to throw around. In particular, we should say what it means to exist in a world and to have a property in a world. An individual is said to exist in a world just in case it is impossible for that world to be actual and that individual to fail to exist. Alternatively, we could say that the individual would exist, were the world in question actual. An individual has a property in a world if and only if it is impossible for that world to be actual while the individual lacks the property.

Now we are ready to try to give a properly *de re* reading of statements like Socrates is essentially rational. An individual is said to have a given property essentially if and only if it is not possible that the individual exist and lack the property. Alternatively, an individual must have the property in every world in which it exists. This account lacks the counterintuitive implications of the proposal just considered above, but it has some implications of its own that should be observed. The first implication is that anything that exists has existence essentially. This may seem implausible initially, but it does accord with the fact that philosophers have found existence to be a peculiar property at best, so peculiar that many philosophers have been led to declare that existence is not a property.

We can deal with possible properties in a similar way. Socrates is possibly Roman-nosed if in some possible world in which Socrates exists, he has the property of being Roman-nosed.

There is another assumption usually made with these accounts of essential properties and possible properties.[29] It is that if individuals — contingent individuals like Socrates — exist in exactly one possible world, then all their properties would be essential to them, their essential properties would be the same as their possible properties, and they would have no accidental properties. Once again, this is a view that runs counter to our modal intuitions.

There are modal theories that have all "world-bound" individuals — individuals who exist in one and only one possible world. This sort of theory was held by Leibniz. An interesting way of preserving our modal beliefs while holding that all individuals are "world-bound" is to have counterparts in other worlds for a given individual in a given world. Leibniz sometimes seems to have thought of matters this way (see his *Correspondence with Arnauld*). Recently David Lewis has worked out a theory of this sort.[30]

I prefer to stick with what I think is our ordinary view. Individuals exist in some worlds and not in others. They are not "world-bound" and the set of individuals can vary from world to world. Some possible worlds lack individuals who exist in the actual world and some possible worlds contain individuals who do not actually exist.

There are at least two accusations one might be tempted to bring at this point. One is the charge that we are now committed to holding that there are things that do not exist. But all we are committed to saying is that there are sets of properties that would have been exemplified if some other world had been actual. The same sort of answer can be given to the charge that we are committed to the existence of Santa Claus, Superman, Captain Nemo, and Hamlet. Indeed, these seem to have some status in the actual world. But we may reply that such fictional characters, although they may even bear names, are not individuals but sets of properties that may or may not describe individuals.[31]

4.6 Considerations Bearing on Quantified Modal Logic

Our informal look at possible worlds has generated several considerations that will have some bearing on what we will accept as a correct system of quantified modal logic. Of course, when we speak of a "correct" system, we mean one that is correct relative to a certain interpretation. So our look at possible worlds, essential properties, and so forth shows us certain things that our system, under the intended interpretation, should either make turn out true or at least should not rule out.

Here is a list of the considerations that have emerged so far:

1. Only one world is actual.
2. The only actually existing individuals are the individuals who exist in the actual world.
3. Individuals exist in more than one possible world.
4. There can be individuals existing in some possible worlds who do not exist in the actual world.
5. There are individuals existing in the actual world who fail to exist in other possible worlds.
6. Individuals have some of their properties essentially and other properties accidentally.
7. An individual has a property essentially just in case that individual has that property in every world in which that individual exists.
8. The *de re* understanding of 'necessarily' or '□' is as '*essentially*'.
9. Individuals do not have properties in worlds in which they do not exist.
10. There are names of individuals that refer to that individual in every world in which that individual exists (rigid designators).

Most of these considerations have to do with existence, and there is a good reason for that. We are introducing quantified modal logic by adding quantification to propositional modal logic. On the usual understanding of the quantifiers, quantifiers range over all and only existing individuals. A universally quantified formula is true just exactly when every actually existing individual satisfies (makes true) the formula preceded by the universal quantifier, and an existentially quantified formula is true just in case some existing individual satisfies (makes true) the formula preceded by that existentially quantified formula. We want to preserve this understanding when we introduce modality; we do not want our quantifiers to be understood as ranging over nonexisting things. Quantifiers that occur within the scope of modal signs, that is to say, quantifiers that occur within *de dicto* contexts, receive no special reading. Quantifiers into *de re* modal contexts are understood to be ranging over actually existing individuals. Perhaps some examples will be helpful.

(A) $\Box$ $(x)(x$ is human $\supset x$ is rational) is to be understood as saying that in each possible world it is true that all humans
are rational.

(B) $(\exists x)(x)$ is a human & $\Diamond(x$ high-jumps 8 feet)) is understood as saying that there is an actually existing individ-
ual who is human and in some possible world that individual high-jumps 8 feet.

Some of the considerations we listed above are reasons to preserve
this understanding of quantifiers in quantified modal logic.

4.7   The Systems

We turn at last to some systems of quantified modal logic. The
proposals we will consider are all of fairly recent origin, reflecting the
fact that only recently have serious advances been made in this branch
of logic, despite the long history of the modal concepts. We will be
looking at quantified S5, although anyone who wants to deal only
with quantified T or quantified S4 can so restrict things by allowing
use only of T-reit or of S4-reit, depending on the system desired. The
quantification rules and other restrictions remain the same from system to system.

4.8   A Kripke System (1959)

The simplest and most straightforward way for us to generate a
system of quantified modal logic at this point is to add the quantifier rules given in chapter III to the modal logic and propositional
logic rules we already have. All we need do is extend the definition
of a well-formed formula so that predicates (i.e., expressions containing free variables) are also governed by our operators and subject to
our rules. In effect we have already presupposed an understanding
of this.

The resulting system of quantified modal logic is equivalent to one
proposed and studied by Saul Kripke in 1959, and by Hughes and
Cresswell (1968). Kripke took his system to be quantified S5. Later
he revised that claim, for reasons we will consider shortly. This Kripke
system is easy to use, requires no special restrictions, and presents
no special technical problems. It is simply the combination of the
methods of the first three chapters. However, it is crucially impor-

tant to note one thing. The *nec intro subproof* rule uses the reiterated lines as assumptions. Any line introduced into a nec intro subproof (this includes poss elim subproofs) by a reiteration rule (T-reit, S4-reit, S5-reit) is an assumption within that subproof and thus is subject to restriction (3) in the statement of the Universal Generalization rule. In other words, if we bring a formula containing a free variable into a nec intro or poss elim subproof by means of a reiteration rule, we may not apply UG to that variable within that subproof. To illustrate this system, we will prove a few theorems.

First, we prove a quantified *de re* version of the characteristic formula of S5:

$(x) \lozenge Fx \supset (x) \square \lozenge Fx$
$(x) \lozenge Fx$	asp
$\lozenge Fx$	UI
$\square \mid \lozenge Fx$	S5-reit
$\square \lozenge Fx$	nec intro
$(x) \square \lozenge Fx$	UG
$(x) \lozenge Fx \supset (x) \square \lozenge Fx$ impl intro

The following more interesting formula is also provable in this system:

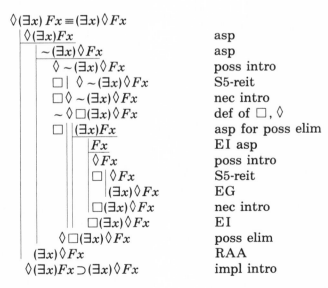

$\lozenge (\exists x) Fx \equiv (\exists x) \lozenge Fx$
$\lozenge (\exists x) Fx$	asp
$\sim (\exists x) \lozenge Fx$	asp
$\lozenge \sim (\exists x) \lozenge Fx$	poss intro
$\square \mid \lozenge \sim (\exists x) \lozenge Fx$	S5-reit
$\square \lozenge \sim (\exists x) \lozenge Fx$	nec intro
$\sim \lozenge \square (\exists x) \lozenge Fx$	def of $\square$, $\lozenge$
$\square \mid \mid (\exists x) Fx$	asp for poss elim
Fx	EI asp
$\lozenge Fx$	poss intro
$\square \mid \lozenge Fx$	S5-reit
$(\exists x) \lozenge Fx$	EG
$\square (\exists x) \lozenge Fx$	nec intro
$\square (\exists x) \lozenge Fx$	EI
$\lozenge \square (\exists x) \lozenge Fx$	poss elim
$(\exists x) \lozenge Fx$	RAA
$\lozenge (\exists x) Fx \supset (\exists x) \lozenge Fx$ impl intro

and

$(\exists x)\Diamond Fx$	asp
$\sim\Diamond(\exists x)\,Fx$	asp
$\Box(x)\sim Fx$	Def, DN
$\Box\,(x)\sim Fx$	T-reit
$\sim Fx$	UI
$\Box\,\sim Fx$	nec intro
$(x)\Box\,\sim Fx$	UG
$\sim(\exists x)\Diamond Fx$	Def, DN
$\Diamond(\exists x)\,Fx$	RAA
$(\exists x)\Diamond Fx \supset \Diamond(\exists x)\,Fx$	impl intro

Exercises

1. $(x)\,(Fx \rightarrow Gx) \supset ((x)(Gx \rightarrow Hx) \supset (x)\,(Fx \rightarrow Hx))$
2. $(\Box(x)(Fx \supset Gx)\ \&\ \sim(\exists x)\Diamond Gx) \supset (x)\Box \sim Fx$
3. $\Diamond(x)Fx \supset (x)\,\Diamond Fx$
4.   Try to explain why the converse of (3), $(x)\Diamond Fx \supset \Diamond(x)Fx$, should *not* be valid. Try to think of a counterexample. What restriction on our rules blocks the attempt to prove it?
5. $(x)(Fx \rightarrow Gx) \supset ((x)Fx \rightarrow (x)\,Gx)$

Alas, mixing quantification and modality is not as easy as this system makes it appear. The example and exercises you have just finished are designed to illustrate where the difficulties are. The difficulties philosophers have had with this Kripke system center on the logical relationships between *de dicto* and *de re* modalities. There are four notorious formulas which are the focal points of controversies:

Barcan Formula:[32]	$(x)\Box \mathfrak{a}x \supset \Box(x)\mathfrak{a}x$
Converse Barcan Formula:	$\Box(x)\mathfrak{a}x \supset (x)\Box \mathfrak{a}x$
Buridan Formula:[33]	$\Diamond(x)\mathfrak{a}x \supset (x)\Diamond \mathfrak{a}x$
Converse Buridan Formula:	$(x)\Diamond \mathfrak{a}x \supset \Diamond(x)\mathfrak{a}x$

These four formulas, as we are interpreting them, express relations of modality *de dicto* to modality *de re*. The Barcan Formula, if part of a system of modal logic, would sanction the inference of a *de dicto* necessity from a *de re* counterpart, and the Converse Barcan Formula licenses the reverse inference. The Buridan and Converse Buridan Formulas would do the same for *de dicto* and *de re* possibility. As anyone who has read the text and done the exercises now knows, the system under consideration has as theorems the Barcan, Converse Barcan, and Buridan formulas. It also is easy to see that the corresponding versions of quantified T and quantified S4 also contain the Buridan Formula and the Converse Barcan Formula, but not the Barcan Formula, since the proof of the latter requires the use of *S5 reiteration*. The Barcan Formula can be added to these systems without generating any formal inconsistency, however.

4.9   Objections to This System

We have observed earlier that objections to systems of modal logic are typically not merely formal objections. They are objections to a certain interpretation of a formal system. The objections considered here are also of this latter sort. These objections presuppose that we are trying to present a certain type of possible worlds interpretation. The objections are ways in which the present formal system, under interpretation, fails to represent the truth of certain modal claims.

The objections to this first attempt to add quantification to the S5 system of modal logic can be seen clearly if we pursue two related questions: 1) How does our view of possible worlds fit with this system? and 2) what is the relationship between *de re* and *de dicto* modality in this system?

We start with the second question. The answer is provided for the most part by the presence in the system of certain formulas, formulas that relate the representations of modality *de dicto* and modality *de re* under the interpretation we are considering. We have in this system as a theorem, the Barcan Formula

$$(x)\Box ax \supset \Box(x)ax,$$

the Converse Barcan Formula

$$\Box(x)ax \supset (x)\Box ax,$$

and the Buridan Formula

$$\Diamond(x)ax \supset (x)\Diamond ax.$$

The Barcan Formula and its converse make *de dicto* and *de re* necessity equivalent in this system. The Buridan Formula allows us to infer *de re* possibility from *de dicto* possibility. We have already noted that its converse is not a theorem of this system.

The Barcan and Buridan Formulas have proven to be highly controversial under our present interpretation. Indeed, the situation is even worse than that. These formulas seem to express false claims under our interpretation. Counterexamples to both may be found in Plantinga's *The Nature of Necessity* (Oxford, 1974), chap. 4, §9.

Plantinga uses a counterexample to the Buridan Formula given by Buridan himself in the 14th century. It seems to be possible that God should not have created anything and that therefore it is possible that nothing exists besides God. But it does not follow from this that everything that in fact exists is possibly identical with God. For example, you and I, who are actually existing things, are not even possibly identical with God. There is no possible world in which I am God and, for that matter, none in which you are either.

Likewise, the Barcan formula does not express a necessary truth. Imagine a universe in which the only existing things are *essentially* immaterial objects – God, sets, numbers, properties, etc. Then it would be true that everything is essentially immaterial, but it would not follow that it is a necessary truth that everything is an immaterial object, since God could (and did) create some material objects.

On the other hand, it seems correct that the Converse Barcan formula should be included as a theorem of our system of modal logic, and also correct that the Converse Buridan Formula should be excluded. Concocting a counterexample to the latter is left as an exercise for the reader.

Other difficulties occasioned by the Barcan and Buridan formulas show up when we consider what sort of possible worlds are required to satisfy the present system of quantified modal logic. The counterexamples make use of sets of possible worlds with different individuals in them. But any possible worlds model of our system will have to have the same individuals in every possible world.

To see this, suppose that we have a possible world in which everything has a certain property a essentially. That is to say, everything in this world has a and has a in every world in which it exists. As we have just seen from the counterexample to the Barcan Formula, we cannot count on everything in another world having a if there are individuals existing in the latter world who do not exist in the former. Having the Barcan Formula valid means in effect that no world can have individuals in it who do not exist in every other world. Since the actual world is the standard for the things there are, the Barcan Formula, if valid, would have the effect of making it the case that no possible world has any individuals who do not exist in the actual world, a strongly counterintuitive conclusion.

This conclusion can be evaded, but only at the price of other implausibilities. One way to evade this conclusion is to hold that things can have properties in worlds in which they do not exist. Another is to say that all essential properties are the kind of properties nothing can fail to have in any possible world – properties like being either red or non-red and like being such that $2 + 2 = 4$. These sorts of properties are sometimes called trivially essential properties. So to have a quantified modal logic that can represent significantly essential properties and do so without requiring that individuals have properties in worlds in which they do not exist or requiring that no possible world have any individuals that do not exist in the actual world, we will have to jettison the Barcan Formula.

We should consider, too, what effect including the Buridan Formula has on our possible worlds model. By the Buridan Formula, if there actually is something that has an essential property, then in every world there is something that has that property. Now if each individual is essentially unique, i.e., has some property essentially that cannot be possessed by any other individual, then no individual in the actual world can be absent from any other possible world. For example, if being identical with Socrates is essential to Socrates, and only Socrates can have this property, as seems plausible, then there are properties of this sort. But then we have the conclusion that no world lacks any individual who exists in the actual world.

Combining the results of considering these two formulas, we have the conclusion that exactly the same individuals exist in all possible worlds. But that implies, of course, that all individuals exist necessarily. So either we are stuck with that blatant falsehood, or we have a modal system that is satisfied only by necessary beings – those things that exist necessarily. This means that the system cannot deal with essential properties of contingent beings, such as you and me and our earthly possessions. Either way it seems desirable to look for another system.

4.10   Kripke's Revised S5 (1963)

The problem is this. Just a few years before Kripke's 1959 paper, Arthur Prior had published a deduction of the Barcan formula from the axioms of standard quantification theory and S5.[34] It seemed therefore that the Barcan Formula (and the Buridan and Converse Barcan Formulas) were simply part of quantified S5 and that it was impossible to get rid of them without going to a different system. But in 1963 Kripke claimed that there is a subtle fallacy in Prior's proof and that the system we have just considered is an improper version of S5.

In presentations of quantification theory, one of the axioms is customarily stated as follows:

$$(x)\, \alpha x \supset \alpha y.$$

Strictly speaking, it is an open sentence, not a proposition, as it is here stated. Of course, Kripke pointed out, it is understood to mean

$$(y)\,((x)\alpha x \supset \alpha y)$$

and ordinarily not writing down that initial quantifier creates no problems. However, Prior's derivation of the Barcan Formula requires applying the rule of necessitation (If p is a theorem, then $\Box p$ is a theorem) to the formulation of the axiom with the initial quantifier dropped. The proof will not work if we may use only the second, stricter form of the axiom. What I have called the stricter form of the axiom is also referred to as the *universal closure* of the earlier form. A universal closure of a formula is obtained by binding all free occurrences of variables in the formula with universal quantifiers prefixed to the formula.

Kripke's "revision" of S5 involves no different formulations of the axioms, just a change in the way quantification is understood to apply to them. We now understand only the universal closures of these formulas to be axioms. Kripke's contention is that this should have been the proper understanding all along. The effect of this understanding is that neither the Barcan Formula, the Converse Barcan Formula, nor the Buridan Formula are theorems of quantified S5. They may be added without generating inconsistency, but they are not part of the system.

We can obtain this "revised" system with our natural deduction rules by adding certain restrictions to the necessity introduction rule and to the possibility elimination rule.

The necessity introduction rule
(for revised S5 quantified modal logic)

Recall that this rule provides a way of terminating a nec intro subproof. The nec intro subproof rule remains as it was. The additional restrictions introduced here will only affect the derivation of certain formulas containing free variables and formulas obtained by a generalization rule. Propositional logic will not be affected at all.

nec intro

A nec intro subproof may be terminated at any point, and any line, p, occurring within the subproof and that is not within the scope of an undischarged assumption within the subproof, may be entered as a line prefixed by □ with the justification "nec intro," provided that (i) the derivation of this line, p, does not depend on the application of UG or EG to a formula containing a free occurrence of a variable which also has a free occurrence in a line preceding the opening of the nec intro subproof in question, and provided that (ii) p contains no free occurrence of a variable which has been introduced into the nec intro subproof by UI or EI. (The mention of EI here is, strictly speaking, unnecessary, because the restrictions on the EI rule will also block the cases involving EI that the restriction is designed to forbid.)

The first proviso of this revised rule uses a technical concept of "depends on," which is defined as follows.

> A line l_j of a proof (or subproof) *depends on* a line l_i iff (i) $l_i \supset l_j$ or (ii) l_j is justified as a direct consequence of rules or equivalences from some premises, at least one of which depends on l_i.

The general idea is that if a line (a formula) is used in the deduction of another line in a proof, the latter depends on the former.

Possibility elimination (revised quantified modal logic)

The rule of possibility elimination also needs to be restricted because, as we have seen earlier, it is officially an abbreviation of a special case of nec intro.

poss elim

> Where $\Diamond p$ appears as a line in a proof (and does not lie within the scope of a discharged assumption), and where an impl intro subproof opened with p as its assumption within the scope of a nec intro subproof yields a conclusion q, both subproofs may be simultaneously terminated and $\Diamond q$ entered as the next line with the justification poss elim, provided that (i) neither q nor any line on which q depends results from the application of UG or EG to a formula containing a free occurrence of a variable that has a free occurrence preceding the opening of the nec intro subproof in question, and provided that (ii) q contains no free occurrence of a variable introduced into the nec intro subproof by UI or EI.

The point of these restrictions or provisos is to prevent interchange of quantifiers and modal signs. This is done by forbidding the application of generalization rules within modal contexts (i.e., nec intro subproofs) to variables having free occurrence outside those contexts (the first proviso), and by forbidding us to bring out of modal contexts free variables that were introduced by instantiation within the modal context (the second proviso).

A good way to test and illustrate these restrictions is to apply them to proofs of the Barcan and Converse Barcan formulas that would be acceptable in the earlier system. Consider this proof of the Converse Barcan formula:

$$\Box(x)\alpha x \supset (x)\Box \alpha x$$

1.	$\Box(x)\alpha x$	asp
2.	$\Box$	$(x)\alpha x$	T-reit
3.		αx	UI
4.		$\Box \alpha x^*$	nec intro
5.		$(x)\Box \alpha x$	UG
6.	$\Box(x)\alpha x \supset (x)\Box \alpha x$		impl intro

*Step 4 now is forbidden by the second proviso, which requires that the step to which nec intro is applied contain no free occurrences of a variable introduced by use of an instantiation rule within the subproof. Line 3 contains a free occurrence of a variable introduced into the nec intro subproof by UI.

As further exercises, look back at the proofs given in Section 9 of this chapter of the Barcan and Converse Barcan formulas and determine which restrictions are now violated in those proofs. It may also be a useful exercise to attempt proofs of the Buridan and Converse Buridan formulas ($\Diamond(x)\alpha x \supset (x)\Diamond \alpha x$ and $(x)\Diamond \alpha x \supset \Diamond(x)\alpha x$) to see how such attempts run afoul of the restrictions.

The system under discussion includes as its theorems all and only the *closed* formulas deducible by means of the natural deduction rules given. A *closed formula* is one that contains no free occurrences of variables. Many formulas containing free occurrences of variables are deducible, but they are not theorems of the system. Kripke says in stipulating this that such open formulas only make sense if we understand them as implicitly closed anyway, so we are not depriving ourselves of anything we should have, and we are getting rid of some undesirable formulas. This system he regards (with some justice) as the proper version of quantified S5. It has the advantage of not yielding the Barcan and Buridan formulas, which run counter to the sense of the modalities we want to capture. The system also has the advantage of not requiring that the same individuals "populate" each possible world. We should remind ourselves here that these advantages are advantages only when we consider the systems under our *intended* interpretation.

4.11   Reflections on Kripke's Revision

This system of quantified S5 presented by Saul Kripke in 1963 is better (for our purposes) than the version found in his 1959 paper. But it is still not fully satisfactory. Our objections will once again be directed against the intended interpretation. While the system as a formal system may be perfectly good for some other applications, here we are interested in its adequacy as the proper structure for the possible worlds picture of modality, including *de re* and *de dicto* relationships.

First, a deficiency. This version of quantified modal logic includes *none* of the principles relating modality *de re* and modality *de dicto*: the Barcan and Converse Barcan formulas as well as the Buridan and Converse Buridan formulas are false under the intended interpretation. We have already endorsed counterexamples to the Barcan and Buridan formulas. We have not yet said anything about the Converse Barcan formula

$$\Box(x)ax \supset (x)\Box ax.$$

Earlier it was suggested that the Converse Barcan Formula is acceptable. Yet Kripke offers a counterexample (strictly speaking, a countermodel) to it right along with his counterexample to the Barcan Formula. We need to look further at this to see how his revised system offers a different account of modality than what we are looking for.

Kripke presents a counterexample along the following lines. Suppose there are two possible worlds, the first containing as its only individuals Ronald and Nancy Reagan, and the second containing only Ronald. In the first world each is bipedal, in the second world Ronald is bipedal but Nancy is not, since she does not exist in that world. In each world, however, everything is bipedal. Hence it is true in *every* world (all two of them) that everything is bipedal, that is, it is necessary that everything is bipedal. But it is false that everything is necessarily bipedal because Nancy is not bipedal in the second world. Thus, the Converse Barcan Formula does not express something true in all possible worlds and is not a principle of quantified modal logic.

This counterexample serves to make clear that Kripke's interpretation of □ differs from what we were looking for, and that the absence of the Converse Buridan Formula is a deficiency. Kripke reads the Converse Barcan Formula as saying that if it is true in every possible world that everything in that world has property a, then a is true of everything there is and true of it in every possible world. Since Kripke also assumes, rightly, that things do not have atomic properties in worlds in which they do not exist (his interpretation makes it false in such a world that the individual has the property), his system is incapable of expressing atomic essential properties of entities that do not exist in every possible world. This means that no contingently existing individuals, such as Socrates, can be said to possess essentially such properties as being human or being rational, assuming these to be atomic properties. For it is not true of Socrates in every possible world that he is human or that he is rational. Kripke's system is only capable of expressing trivially essential properties of contingently existing entities by using the □. A trivially essential property is one that every individual has necessarily, such as being red or non-red or, perhaps, the property of being such that $2 + 2 = 4$.

If, on the other hand, we understand having a property essentially as not lacking that property in any world in which it exists, and if we want □ as attached to open sentences to express possessing a property essentially, then it can be seen that the Converse Barcan Formula should be a theorem of our system. For the CBF simply expresses what is a necessary truth: if it is true in every possible world that everything is a, then everything has a in every world in which it exists. And we continue to assume that individuals do not have properties in worlds in which they do not exist.

There is also a metaphysical difficulty with Kripke's interpretation of quantified S5 that we will not pursue in detail here: it seems to require that there are some individuals that do not exist. According to the interpretation he proposes, for each possible world there is a set of individuals that comprises the individuals that exist in that possible world. There will be some "overlap" from world to world, but also some individuals "lost" and others "added." Taking this interpretation, there must be a union of all these sets of individuals — the total set of all individuals in all possible worlds. Now if we suppose, as seems entirely plausible, that the actual world does not contain all the individuals there could be, then some other possible worlds "contain" individuals not to be found in the actual world. But if this is so, then there must be a union of all the sets of individuals in all the possible worlds that contains objects that do not actually exist. Thus, anyone who accepts this kind of interpretation of quantified S5 is committed to the implausible claim that there are some things that do not exist.[35]

4.12   The Actualistic Version of Possible Worlds[36]

Despite the claims in the propositional modal logic section of this book that S5 best reflects the truth about matters modal, both versions of quantified S5 considered here have been metaphysically unsatisfactory. We will see shortly, however, that only a slight modification is required to produce a satisfactory system. But first we should take a closer look at the metaphysical picture of possible worlds the system must fit – the actualist view of possible worlds.

Actualism is a species of what textbooks on metaphysics traditionally call realism. However, actualism is an attempt to avoid the excesses that have sometimes been associated with realism. While propositions, properties, and states of affairs all exist for the Actualist, there are no nonexistent objects, nor could there be any. Furthermore, the "serious actualist" holds that objects do not have properties in worlds in which they do not exist, and he believes it is possible that there should be objects distinct from all the objects there in fact are. The actualist rejects any interpreted modal system attempting to represent broadly logical necessity that fails to honor these truths.

But we should begin at the beginning – possible worlds. What does the actualist think they are? We will explain them in terms of states of affairs. We assume that there are states of affairs. There are present states of affairs like Ronald Reagan's being President, past ones like Roger Maris's hitting his 61st home run, future ones like the Second Coming of Christ; there are permanent states of affairs such as any equilateral triangle being equiangular and 7 plus 5 being 12; there are states of affairs that never have been and never will be actual, such as Napoleon's being very tall; and there are states of affairs that could not be actual, such as the Barber of Seville's shaving all and only those persons in Seville who do not shave themselves. All of these states of affairs exist; but not all of them do, or could, actually obtain.

A possible world is a special state of affairs. It is, of course, first of all a *possible* state of affairs. (Remember we are not trying to *explain* the meaning of possibility here.) Secondly, to be a possible world, a possible state of affairs must be *maximal*. What does it mean for a state of affairs to be maximal?

To understand this, we introduce two other concepts, that of one state of affairs *including* another and that of one state of affairs *precluding* another. Let S_1 and S_2 stand for any states of affairs (not necessarily different ones). Then S_1 is said to *include* S_2 if and only if it is not possible for S_1 to be actual without S_2 also being actual, and S_1 precludes S_2 if and only if it is impossible that S_1 and S_2 should both be actual. So, for example, the state of affairs of Roger Maris's hitting his 61st home run includes the state of affairs of Roger Maris's having hit his 30th home run, while the state of affairs of something being an obtuse triangle precludes that thing's being an equilateral triangle.

These concepts now enable us to define a maximal state of affairs. A state of affairs, *M,* is *maximal* just in case for any state of affairs, *S, M* either includes *S* or precludes *S.*

According to this definition, given a maximal state of affairs and any old state of affairs, the latter state of affairs is either already part of the maximal state of affairs or cannot be added to the maximal one without yielding an inconsistent state of affairs as a result. Any genuine enlargement of a maximal state of affairs will be inconsistent.

Notice, however, that a maximal state of affairs need not be composed of a large number of constituent states of affairs. It may be, but it may also be composed of just one – a self-inconsistent one, such as $7 + 5 = 13$, which will, of course, preclude every state of affairs. On the other hand, however, a check of the definition of *includes* shows that inconsistent states of affairs turn out to *in*clude every state of affairs, too. The point is that any inconsistent state of affairs is maximal.

It is clear that we want nothing to do with these inconsistent maximal states of affairs. We define a possible world, then, as a state of affairs that is both maximal and consistent. This explanation accounts for possible worlds without the need to introduce any shadowy metaphysical entities or strange doctrines. Possible worlds can be nicely explained in terms of entities with which everyone is quite familiar.

Not only is the notion of a possible world thus explained in quite familiar terms, it is also explained in a way that makes it perfectly plausible to suppose that there are possible worlds. If there are states of affairs, as there surely are, then there are possible worlds. It has to be kept in mind, of course, that these are abstract entities. It is not as though there is some *place* at which possible worlds exist. All possible worlds, including the one that has been actualized, are abstract entities. Now, from this it does *not* follow that the *physical* world is an abstract entity, but rather that the existence of a pos-

sible world differs from the *actuality* of a possible world. All of a huge
number of possible worlds exist; at most one can be and is actual.

4.13   Actualism and Quantification

Inasmuch as we have criticized other interpretations of modal logic
for requiring that there are things that do not exist or that things
can have properties in worlds in which they do not exist, we should
see how the actualist system and interpretation handles these problems. We begin by spelling out how quantifiers are to be interpreted
in our system.

One's first inclination is to start with the reading of ordinary quantification theory and extend that kind of interpretation to possible
worlds. Quantification is customarily construed as being over all existing objects.

On our ordinary reading of the quantifiers, there will be a domain
of quantification — a set of objects that are the values of variables.
This domain may be restricted for various purposes, e.g., mathemati-
cians may confine the domain to numbers or sets in the study of
arithmetic or algebra, but ordinarily this domain will consist of everything there is. Of course, philosophers and others notoriously disagree
about what there is, but this is a metaphysical disagreement, not one
about interpreting quantifiers.

Quantified statements are interpreted with reference to the domain. Universally quantified statements are understood as saying
that every member of the domain satisfies the condition that follows
the quantifier. Or, stated less technically, the predicate or property
denoted by the expression following the universal quantifier is true
of everything in the domain. And likewise, following the definition
of the existential quantifier in terms of the universal quantifier, an
existentially quantified statement is understood as claiming that it
is false that everything in the domain fails to satisfy the condition
expressed by the expression following the quantifier.

Now it might seem natural to continue to use this interpretation
when working with quantified modal logic, but we have already hinted
that there are certain difficulties involved with it. The set of objects
existing in a possible world may differ from world to world, and we
want to understand quantification in a possible world to be over the
objects that exist in that world (i.e., over the objects that would exist if that world were actual). However, the only domains of contingent objects actually available are the set of objects that actually
exist and domains that are subsets of that set. If we try to get by

without allowing for objects that exist in other possible worlds but not in the actual world, the alternatives are unpalatable. We either (1) have to explain quantification in a possible world when part or all of its domain does not exist, or (2) explain how something can have a property (at least a nontrivial property) in a world in which it does not exist, or (3) hold that there are things that do not exist, or (4) hold that there are no possible worlds containing objects not contained in the actual world.

Assuming we want to hold to a standard reading of quantifiers and we want some possible worlds to "contain" individuals who do not exist in the actual world, how shall we construe domains of quantification for possible worlds? The domains of quantification for the various possible worlds will have to be subsets of actually existing things and be necessarily existing things in order to satisfy the requirements we have laid down. Alvin Plantinga suggests that we use essences — individual essences.

What are individual essences and how can it be made clear that they will serve as the domain of quantifiers in modal contexts? We have said that an object *has a property in a world* if it would have been true that it has the property had that world been actual. For an object *to have a property essentially* is to have the property in every world in which it exists. The introduction of one further concept will enable us to define individual essences. Let us define what it means for a property to be exemplified in a possible world. A *property is exemplified in a possible world* if the proposition that something has that property would have been true had that world been actual. This definition enables us to offer our official definition of an individual essence. An individual essence is a property that is exemplified in some possible world, and in every possible world any thing that has this property has it essentially, and nothing else in any possible world has this property.[37] Although a given individual essence need not be exemplified in any particular possible world, essences are properties, and since properties presumably are necessary beings, as such, they exist in every possible world.[38] So, for example, Julius Caesar's essence exists in every possible world even though Julius Caesar fails to exist in some possible worlds. We can exploit this feature of essences to explain quantification in modal contexts.

To conclude our explanation, we need to define one more concept, that of coexemplification. We will say that two properties are coexemplified with each other in a possible world when the proposition that something has both properties would have been true had that possible world been actual.

With this, we are in a position to say how the actualist who adopts quantification over domains of essences reads various types of propositions. Consider first of all ordinary universally quantified propositions and existentially quantified ones.

$$(x) (x \text{ is a man} \supset x \text{ is mortal})$$

is understood as follows: any essence in the domain of essences that are actually exemplified (i.e., any essence in the domain of the actual world) is such that if it is coexemplified with the property of being a man, then it is coexemplified with the property of being mortal. The existentially quantified statement

$$(\exists x) (x \text{ is man} \ \& \ x \text{ is mortal})$$

is read like this: Some essence in the domain of essences exemplified in the actual world is coexemplified in the actual world with the properties of being a man and being mortal. Some existing thing, which of course has an individual essence and which is an example of (exemplifies) that essence, has the properties of being human and mortal.

The advantages of this actualist view begin to emerge more clearly as we consider statements which mix modalities and quantifiers. First we look at quantifiers within the scope of a modality:

$$\Diamond(x) (x \text{ is a man} \supset x \text{ is mortal})$$

is interpreted as meaning that in some possible world, every individual essence in the domain of essences that would be exemplified if that world were actual is coexemplified with the property of being mortal if a man. Or, more simply, $\Diamond(x)ax$ means that in some possible world, every individual essence that would be exemplified if that world were actual is coexemplified with property a. Note that we say "would be exemplified" and not "would exist." Every individual essence *exists* in every possible world. But not all are exemplified in every possible world. Similarly, $\Diamond(\exists x)ax$ is interpreted as saying that in some possible world, some individual essence (in the domain of essences of that world) is coexemplified with the property a. This reading boils down to saying that $\Diamond p$ is true just in case there is a possible world in which p is true.

How about examples of modalities within the scope of quantifiers, as in

$$(x)\Diamond(ax)$$

and

$$(\exists x)\Diamond \, ax?$$

Here we are no longer dealing with the truth of a proposition in a possible world – the modality of a *dictum* (i.e., modality *de dicto*) – but with claims about objects of the actual world (essences exemplified in the actual world) that assert that these essences are in some (or all) possible world(s) coexemplified with a certain property or properties. $(x)\Diamond\alpha x$ says that every essence in the domain of the actual world (those essences exemplified in the actual world) is such that in some possible world that essence is coexemplified with the property α. The second, $(\exists x)\Diamond\alpha x$, says that some essence exemplified in the actual world is in some possible world coexemplified with the property α.

Now consider a related statement that embeds the modality deeper:

> Some man is possibly mortal, i.e.,
> $(\exists x)$ (x is a man & $\Diamond(x$ is mortal)).

This is read as saying there is an essence in the domain of the actual world, coexemplified with the property of being a man, and this essence is coexemplified in some possible world with the property of being mortal.

Exercises

A. Symbolize and interpret the following. Where the statement is ambiguous, give the possible readings and try to say which is more plausible.
1. Possibly (all) minds can exist apart from (all) bodies.
2. (All) minds can exist apart from bodies.
3. Possibly there is a golden mountain.
4. There is something that could be a golden mountain.
5. All men are rational essentially.
6. All bachelors are essentially unmarried.

B. Argue in terms of the present interpretation for your answers to the following:
1. Does
   a) $(\exists x)\Diamond(x$ is $F) \rightarrow \Diamond(\exists x)(x$ is $F)$?
   b) $\Diamond(\exists x)(x$ is $F) \rightarrow (\exists x)\Diamond(x$ is $F)$?
2. Is $\Diamond(\exists x)(x$ is a flying horse) necessary or contingent?

4.14   Actualistic Quantified Modal Logic

Despite our claims in the propositional modal logic section of this book that S5 best reflects the truth about matters modal, the first two versions of quantified S5 considered here have proven metaphysically unsatisfactory. The satisfactory metaphysical view we have called actualism – there neither are nor could be any nonexistent objects. We also hold that something must exist in a world to have properties in a world, and we hold that other possible worlds have things existing in them that do not exist in the actual world. Now we want a system that under the interpretation outlined in the previous section yields or permits these conclusions and at the same time enables us to demonstrate the validity of standard modal inferences.

The actualist system of quantified modal logic for which we give natural deduction rules was axiomatized by Thomas L. Jager.[39] It is very close to the Kripke system, differing only with respect to some theorems of quantified modal logic. There are actualist versions of T, S4, and S5, and it is a very simple matter to move from one system to the other, just as it was in propositional modal logic. Jager's system A is a quantified modal logic obtained by adding quantification in a certain way to S5 propositional modal logic; actualist versions of T and S4 can be obtained in the same way.

The differences between this actualistic system and the other two systems presented here can be summarized in terms of what they do with the formulas governing the mixing of quantification and modality – the Barcan and Converse Barcan formulas, and the Buridan formula. (All three systems exclude the Converse Buridan formula.) The first Kripke version of S5 had all three, and the second rejected all three. The actualistic system excludes the Barcan and Buridan formulas but includes the Converse Barcan formula as a theorem.

4.14.a   Rules

We obtain the actualistic system of quantified modal logic by taking our rules for propositional modal logic, adding quantification rules, and restricting the use of the rules *nec intro* and *poss elim*. The restrictions will apply to these rules only in contexts involving quantifiers and free occurrences of variables. The restriction on the *nec intro* rule is designed to prevent us from "losing" variables that have a free occurrence in lines used in deducing a conclusion to which we want to apply the *nec intro* rule. The restriction on the *poss elim* rule is designed to prevent us from "gaining" free occurrences of variables between the point of making an assumption for the subproof used in the rule and the use of the *poss elim* rule.

Actualistic quantified modal logic then, has the following rules:

Nec elim
Poss intro
Nec intro subproof rule
Reit (T-reit, S4-reit, S5-reit)
Quantification rules (UI, UG, EI, EG)
Nec intro$_A$

A nec intro subproof may be terminated at any point, and any line, p, occurring within the subproof and that is not within the scope of an assumption undischarged at that point in the nec intro subproof may be entered as a line prefixed by □ with the justification "nec intro," *provided that the line* p *contains a free occurrence of every variable that has a free occurrence in any line within the nec intro subproof on which* p *depends.*

Poss elim$_A$

Where $\Diamond p$ appears as a line in a proof (and not within the scope of a discharged assumption) and where an impl intro subproof opened with p as its assumption within the scope of an nec intro subproof yields a conclusion, q, both subproofs maybe terminated simultaneously and $\Diamond q$ entered as the next line in the proof with the justification *poss elim, provided that every variable which has a free occurrence in* q *also has a free occurrence in* p.

The best way to see how the restrictions work and how to use them is to consider some examples. Consider the following proof and be prepared to compare it with the succeeding one:

Prove: $(x) \Box ((Fx \& Gx) \supset Hx) \supset (x)(\Box (Fx \& Gx) \supset \Box Hx)$

1.	$(x) \Box ((Fx \& Gx) \supset Hx)$	asp
2.	$\Box (Fx \& Gx)$	asp
3.	$\Box ((Fx \& Gx) \supset Hx)$	UI
4.	$\Box \mid Fx \& Gx$	T-reit
5.	$(Fx \& Gx) \supset Hx$	T-reit
6.	Hx	MP
7.	$\Box Hx$	nec intro
8.	$\Box (Fx \& Gx) \supset \Box Hx$	impl intro 2–7
9.	$(x)(\Box (Fx \& Gx) \supset \Box Hx)$	8,UG
10.	$(x)\Box((Fx\&Gx)\supset Hx)\supset(x)(\Box(Fx\&Gx)\supset\Box Hx)$	impl intro 1–9

Here the only step deduced *within* the nec intro subproof is line 6, derived from lines 4 and 5. Lines 4 and 5 contain free occurrences of the variable x, and x occurs free in line 6, thus satisfying the restriction. Compare this with a similar formula for which a similar "proof" does not go through because of the restriction:

Prove: $(x)(y) \square ((Fx \& Gy) \supset Hx) \supset (x)(y)(\square (Fx \& Gy) \supset \square Hx))$

1.	$(x)(y)\square((Fx \& Gy) \supset Hx)$	asp
2.	$\square(Fx \& Gy)$	asp
3.	$\square((Fx \& Gy) \supset Hx)$	UI
4.	$\square\ Fx \& Gy$	T-reit
5.	$(Fx \& Gy) \supset Hx$	T-reit
*6.	Hx	MP
7.	$\square Hx$	nec intro
8.	$\square(Fx \& Gy) \supset \square Hx$	impl intro
9.	$(x)(y)(\square(Fx \& Gy) \supset \square Hx)$	8 UG
10.	$(x)(y) \square ((Fx \& Gy) \supset Hx) \supset$ $(x)(y)(\square(Fx \& Gy) \supset \square Hx)$	impl intro

The proofs may seem identical, but this latter proof violates the restriction on nec intro subproofs in line 6, the starred line. Line 6 depends on lines 4 and 5. Lines 4 and 5 each contain free occurrences of two variables, x and y, while line 6 contains only an occurrence of x. The free variable y has been "lost." Hence the restriction has not been satisfied and this second "proof" is fallacious.

Next, consider a pair of proofs illustrating the use and misuse of the *poss elim* rule:

Prove: $(x)(\lozenge Ax \supset \lozenge (Ax \vee Bx))$

1.	$\lozenge(Ax)$	asp for impl intro
2.	Ax	asp for poss elim
3.	$Ax \vee Bx$	add
4.	$\lozenge(Ax \vee Bx)$	poss elim
5.	$\lozenge Ax \supset \lozenge(Ax \vee Bx)$	impl intro
6.	$(x)(\lozenge Ax \supset \lozenge(Ax \vee Bx))$	UG

When the *poss elim* rule is applied to line 3 to derive line 4, the restriction is satisfied. Only one variable, x, occurs in line 3 and it is the same one that occurs in line 2, the assumption opening the *poss elim* subproof. Hence, no "new" variables were "gained" within the subproof.

Compare the above proof with the one below. The one below, like the preceding, appears to prove a formula corresponding to a theorem of quantification theory.

Prove: $(y)(x)(\Diamond Ax \supset \Diamond(Ax \lor By))$

1.	$\Diamond Ax$	asp for impl intro
2.	$\Box$ Ax	asp for poss elim
3.	$Ax \lor By$	add
*4.	$\Diamond(Ax \lor By)$	poss elim – violates restriction
5.	$\Diamond Ax \supset \Diamond(Ax \lor By)$	impl intro
6.	$(y)(x)(\Diamond Ax \supset \Diamond(Ax \lor By))$	UG (twice)

Although this proof parallels its predecessor, this latter proof contains a violation of the restriction on *poss elim* in line 4. This time there is a free occurrence of the variable y that does not occur free in line 2, the assumption for the application of *poss elim*.

What happens now to the most exciting formulas such as the Barcan and Converse Barcan formulas? The Converse Barcan formula is easily proven:

Prove: $\Box(x)Fx \supset (x)\Box Fx$

1.	$\Box(x)Fx$	asp
2.	$\Box$ $(x)Fx$	T-reit
3.	Fx	UI
4.	$\Box Fx$	nec intro
5.	$(x)\Box Fx$	UG
6.	$\Box(x)Fx \supset (x)\Box Fx$	impl intro

The restriction on nec intro subproofs is trivially satisfied since the only deduction within such a subproof is the deduction of line 3 from 2, and line 2 contains no free occurrences of any variable. Attempts to prove the Barcan Formula run afoul of restrictions, however. The reader can verify this by turning back to the proof of this formula in section 9, earlier in this chapter. (There the formula appears in an equivalent form, $\Diamond(\exists x)Fx \supset (\exists x)\Diamond Fx$.) The Barcan Formula is not a theorem of this system, which is just what we intended.

4.14.b   Exercises

A. Symbolize
1. There is a being that possibly fails to exist.
2. Possibly there is something brown that could be blue.
3. Necessarily everything that is red is colored.
4. Everything that is red is necessarily colored.
5. Everything that is red is necessarily colored essentially.
6. Anything that is six feet tall might not have been six feet tall.

7. Everything that is essentially rational is visible.
8. Some things that are essentially rational are not essentially embodied.

B. Prove:
9. $(x)\Box((Fx \& Gx) \supset Fx)$
10. $(x)(y)(\Box(Fx \& Gy) \supset Fx)$
11. $(x)\Box((Fx \& Lx) \supset Fx)$

C. Give Proofs:
12. $\Box(x)((Ax \& Bx) \supset \sim Cx)$
    $(x)\Box Ax$
    $(x)\Box(Bx \supset Cx)$ $\therefore (x) \sim \Diamond Bx$

13. $(x)(\Box Ax \supset \Box Bx)$
    $\Box(x)(Bx \supset Cx)$
    $\therefore (x)(\Box Ax \supset \Box Cx)$

14. $(y)\Box((x)Ax \supset By)$
    $\therefore \Box(x)Ax \supset (y)\Box By$

15. $(x)(y)\Box(Gxy \supset Fxy)$
    $(x)(y)\Box(Fxy \supset Cy)$
    $(x)\Box(Cx \supset Bx)$
    $(x)(y)\Box Gxy$ $\therefore (y)\Box By$

D. Attempt to construct proofs of the Buridan and Converse Buridan formulas, and observe which restrictions on our rules prevent the proofs from going through.

4.15   Existence and Modality

Metaphysics abounds with puzzles about existence, possible existence, and necessary existence. Since the interpretation we have adopted has interesting implications for some of these puzzles, it will be worthwhile to take a brief look at a few here.

We have said earlier that any property a thing has in every possible world in which it exists is an *essential property* of the thing. Paraphrasing that in the terms we have adopted, we would say that something has a certain property essentially just in case the essence of that thing is coexemplified with that property in every possible world whose domain contains that essence. It is easy to see that, whichever characterization we use, existence (assuming it is a property) is an essential property of everything that exists. This may seem highly paradoxical at first, since it seems well-nigh axiomatic that many of the things that exist exist contingently. In order to remove this air of paradox, we shall have to make some further distinctions.

Necessity is usually opposed to contingency, while essences or essential properties are usually contrasted with accidents or accidental properties. An accident is a property a thing has but that it could lack and still be what it is. Given this characterization, it seems plausible to deny that existence is an accidental property of any thing that has it. Yet the existence of a good many things, ourselves included, seems to us contingent. Suppose then we call a being *contingent* if it exists (i.e., its essence is exemplified) in some worlds but not in others, while a being is necessary or has necessary existence just in case it exists in every possible world. The air of paradox is dispelled, or at least greatly diminished, when we see that we exist essentially but not necessarily. You exist, and therefore you exist essentially, but it does not follow that you exist necessarily.

We may also use the *de re − de dicto* distinction to explain the difference between necessary and essential existence. What was just called necessary existence can be explained as the truth of a proposition asserting existence of the thing in question in every possible world. Here necessity is predicated of a proposition ascribing existence to the thing. In the case of essential existence, however, the property of existence is predicated of the thing in question, and the thing is said to have this property essentially.

This in turn sheds some light on negative existential propositions, propositions that deny the existence of an individual. The classical paradox holds that it is impossible for negative singular existentials to be true because our reference implies the existence of that whose existence we are denying. This might seem to be a special problem for actualists. By saying that my older brother does not exist, I appear to be asserting something about my older brother. But according to the actualist we cannot predicate properties of things that do not exist. Hence, my older brother has to exist for me to deny that he exists, and so what I am trying to affirm implies its own falsehood.

We have just seen a way of dealing with this. Someone affirming a negative existential may be denying the truth of a proposition or she may be attributing a property, nonexistence, to something. Anyone doing the latter is trying to affirm the impossible. There is no possible world in which such a *de re* claim is true. However, the former sort of claim can truly be made in any world in which the essence referred to by the name is not coexemplified with existence in that world. It is foolishness to try to say of my older brother that he lacks existence; that will always be necessarily false. But it is sensible, and even true, to deny that my older brother exists, i.e., to deny that I have an older brother.

4.16   The Problem of Transworld Identity

The system and interpretation of modal logic with which we have been working holds that some individuals exist in different possible worlds. Indeed, that feature enables us to distinguish a thing's accidental from its essential properties. But we perhaps should pause a moment to acknowledge that not all proponents of modal logic and possible worlds have been happy with that idea. One of the wellsprings of the possible worlds tradition, Leibniz, apparently thought that any given individual exists in one and only one possible world. Indeed, he suggests that God organizes all the possible individuals into mutually compatible (compossible, he says) groups and each individual fits into exactly one. In spite of this, however, Leibniz seems to feel quite free in his correspondence with Arnauld to speak of Adam in other possible worlds, although he will also allow that, strictly speaking, this other Adam is actually a counterpart, and not the same individual (or individual concept).[40] At any rate, Leibniz seems to subscribe to the view that an individual is to be found in at most one possible world. This has been explored more recently by David Lewis.[41]

What are the attractions of such a view? Leibniz apparently believes that all of a thing's properties are essential to it and that to hold less was to impugn the omnipotence and omniscience of God (*Correspondence with Arnauld*, chap. IX). His arguments require a more careful look at his entire metaphysics, and so we shall not pursue them here.

There is another argument raised recently against the view that an individual may exist in more than one possible world. The argument is the allegation that there are insuperable difficulties involved in identifying individuals from one world to the next.[42] A common way of generating the desired sense of bewilderment involves asking us to consider two individuals and to imagine a series of possible worlds in which these individuals exchange more and more properties. Consider Chisholm's example. We begin with Adam and Noah in the actual world. Next we "move" to a possible world in which Adam lives 950 years and Noah lives 930 years. Then we go to a possible world in which Adam lives 950 years and fathers Shem, while Noah lives 930 years and fathers Seth. If we continue this long enough, the argument goes, we will no longer be able to tell who is Adam and who is Noah. There do not seem to be any clear criteria for identifying Adam and identifying Noah such that in any given possible world we can tell which is which. So we are invited to surmise that perhaps there are no individual essences of objects that would make transworld identification possible. Individual essences are supposed to individuate things, and they do not seem to be able to do that.

Initially it is important to note two things about this alleged difficulty. First of all, it should not be taken to imply that there are no individual essences. It does not follow that a thing has no individual essence. Rather, the argument suggests that *all* of a thing's properties are essential to it – a very implausible position. Secondly, the argument or puzzle generates an epistemic claim about our inability (in some cases, anyway) to *tell* what individual essence a thing exemplifies or a claim about our inability to tell whether some thing x in one possible world exemplifies the same individual essence as y in some other possible world. But then it invites us to jump to the metaphysical conclusion that there are no individual essences that can be exemplified in more than one possible world. This conclusion does not follow.

It is also important in this connection not to be misled by the "picture" we have been using. It is helpful to think instead in terms of the official definitions and descriptions given earlier. Recall that possible worlds and individual essences are abstract objects. A possible world is a maximally consistent state of affairs, and an individual essence is a set of properties (or complex property) such that anything that has (exemplifies) it, has (exemplifies) it essentially, and no other thing can have (exemplify) it. Assuming that properties exist necessarily, there is no question about the existence of individual essences, and there is no question that they individuate (i.e., whatever exemplifies one is a distinct individual). There is also no difficulty in the possibility that essences should be exemplified in more than one possible world.

This view leaves open questions about what properties might be essential to various individuals and how we might be able to ascertain whether a given individual has a given property essentially. But this is not apt to confuse us about the nature of essences unless we insist on using the metaphor of visually peering at two incompletely known individuals in two possible worlds and supposing that we should be able to tell whether they have the same essence.

4.17   Superiority of Actualism

One insuperable difficulty of other modal systems we considered was their inability to express the truths

> There could have been an object distinct from every actual object,

and

> There could have been more objects than there are,

without needing to postulate nonexistent objects or ascribing properties to objects in worlds in which they do not exist. The discussion of the previous sections provides the necessary clue for giving readings of these claims without allowing anti-actualist implications.

It will be true that there could have been an object distinct from every actual object just in case there is a possible world in which some essence is exemplified (would be exemplified if that world were actual) that is not exemplified in the actual world. The present interpretation has the virtue of restoring to the original statement its possibility. The companion claim, there could have been more objects than there are, requires that there be more individual essences than are actually exemplified and that an even larger number of them be able to be coexemplified. Again, our interpretation maintains the plausibility of this plausible claim.

Modal actualism thus provides an interpretation of possibility and necessity that preserves and enhances the insights that traditionally have been part of modal logic. It makes clearer the relationships between *de re* and *de dicto* modality and does not quantify into *de dicto* modal contexts. Actualism does not include some of the metaphysical implausibilities of its near competitors and it provides some illuminating insights into the relationship between existence and modality.

APPENDIX I

AXIOMATIZATION OF T, S4, & S5

Although C. I. Lewis was aware that his modal systems contained the propositional calculus, he did not axiomatize his systems by adding modal axioms to an axiomatization of the system of *Principia Mathematica.* Gödel appears to have been first to axiomatize a series of modal systems by adding modal axioms to a complete basis for the classical propositional calculus. Here we follow Gödel.

We start with a complete axiomatization of propositional calculus, taken from Mendelson's *Introduction to Mathematical Logic* (Van Nostrand, 1979), p. 31. The axioms are actually axiom schemata; the letters represent any well-formed formula. Hence, any uniform replacement of letters in the schemata by well-formed formulas results in an axiom. Given the usual conventions about the number of propositional variables available, each axiom schema represents an infinite number of axioms. The propositional calculus base is as follows:

Axioms: 1. $p \supset (q \supset p)$
                  2. $(p \supset (q \supset r)) \supset ((p \supset q) \supset (p \supset r))$
                  3. $(\sim q \supset \sim p) \supset ((\sim q \supset p) \supset q)$
Rule: R1. $p,\ p \supset q / \therefore q$ (modus ponens)

We obtain the system T by adding a definition, a rule, and two axioms:

T = Propositional Calculus (1–3 and R1) plus
                  4. $\Box p \supset p$
                  5. $\Box(p \supset q) \supset (\Box p \supset \Box q)$
Rule: R2. $\vdash p / \therefore \ \vdash \Box p$

Rule R2 says that if a formula p is a theorem (i.e., an axiom or a consequence of axioms), then $\Box p$ is a theorem.

S4 comes by adding its so-called "characteristic formula" as an axiom.

S4: T plus   6. $\Box p \supset \Box \Box p$

119

S5 comes by adding its "characteristic formula" to T.

S5: T plus   7. $\sim\Box p \supset \Box \sim\Box p$

(7.) is the form Gödel gave. It also appears as

　　　　　　　　7'. $\Diamond p \supset \Box \Diamond p$

and　　　　　　7". $\Diamond \Box p \supset \Box p.$

The form (7) is used in order to specify the axiom solely in terms that are primitive, i.e., undefined, within the system.

In an important paper on modal logic, Saul Kripke uses this basis for T as a set of sufficient conditions for calling a modal propositional calculus *normal*. Non-normal systems are ones that do not satisfy R2.

In an interesting paper that groups some modal logics into various "families," entitled "New Foundations for Lewis Modal Systems," E. J. Lemmon defines a *Lewis Modal System* as follows:

A Lewis Modal System is a system that

　(i)　contains the full classical propositional calculus;
　(ii)　is contained in S5;
　(iii)　admits substitutability of tautologous equivalents;
　(iv)　possesses as theorems:
　　　　F1.　$\Box(p \supset q) \supset (\Box p \supset \Box q)$
　　　　F2.　$\Box(p \& q) \equiv (\Box p \& \Box q)$
　　　　F3.　$(\Box(p \supset q) \& \Box(q \supset r)) \supset \Box(p \supset r)$
　　　　F4.　$\Box p \supset \sim\Box\sim p.$

It is easy to see that T, S4, and S5 are Lewis Modal Systems in the defined sense, even though T was not among the systems Lewis himself proposed.

APPENDIX II

PROVING THE EQUIVALENCE OF ALTERNATIVE FORMULATIONS OF T, S4 and S5

We have made a number of references to alternative formulations of the systems we have presented. We have presented each system by means of natural deduction rules and, in Appendix I, by means of axioms. How do we know that we have the system we want? And how do we know that these presentations give the same system?

These systems are usually defined by certain axiomatizations of them. The system T *is* a certain set of theorems, derivable from various specified sets of axioms. The axioms presented for each system in Appendix I are not the original refining ones, but they are sets that yield exactly the same theorems. That different axiomatizations yield the same system can be shown quite straightforwardly.

Suppose we have two sets of axioms and rules, alleged to be axiomatizations of the same system. We must show that whatever is a theorem according to the one axiomatization is also a theorem according to the other. We do this by showing that all the axioms of the one formulation are provable as theorems in the other formulation, and we show how the effect of any rule of the one formulation is obtainable using the rules and axioms of the other. This is, in effect, to show how to turn a proof of a theorem from the one set of axioms and rules into a proof from the other set of axioms and rules. When this can be done, we conclude that the one system *contains* the other.

The procedure is basically the same for proving an axiomatization for a system equivalent to a set of natural deduction rules. We show that the axioms can each be proven via the natural deduction rules, and that the effect of any rules can also be obtained via the natural deduction rules. This shows that the natural deduction system *contains* the axiomatic system. Then we show that the effect of

121

each natural deduction rule can be obtained by means of the axioms and rules of the axiomatic system. Once we show this, we can conclude that the axiomatic system *contains* the natural deduction system. If each contains the other, they are equivalent.

We illustrate this with the modal system T. We will assume here that our natural deduction formulation of the propositional calculus given in chapter 1 and the axiomatization of it (A1–A3 and R1 of Appendix I) are equivalent. We complete the proof for the modal axioms and rules.

A.　Proof that any modal inference obtainable in the natural deduction formulation of T can be obtained from the axioms of T:

Natural deduction rule		*Derivation from axioms of T*	
nec elim			
1. $\Box p$	given	1. $\Box p$	given
2. $\therefore p$	nec elim	2. $\Box p \supset p$	axiom A4
		3. $\therefore p$	MP

nec intro (with T-reit)			
1. $\Box p$	given	1. $\Box p$	given
2. $\Box (p \supset q)$	given	2. $\Box (p \supset q)$	given
3. $\Box\ \vert\ p$	T-reit	3. $\Box(p \supset q) \supset (\Box p \supset \Box q)$	A5
4. $\quad\ \vert\ (p \supset q)$	T-reit	4. $\Box p \supset \Box q$	MP
5. $\quad\ \vert\ q$	MP	5. $\Box q$	MP
6. $\Box q$	nec intro		

B.　Proof that any modal formula that can be shown to be a theorem from the axioms of T can also be derived from the natural deduction rules:

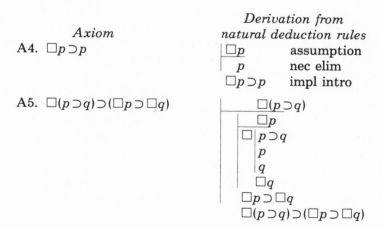

Axiom	*Derivation from natural deduction rules*
A4. $\Box p \supset p$	$\Box p$ assumption
	p nec elim
	$\Box p \supset p$ impl intro
A5. $\Box(p \supset q) \supset (\Box p \supset \Box q)$	$\Box(p \supset q)$
	$\Box p$
	$\Box\ \vert\ p \supset q$
	$\quad\ \vert\ p$
	$\quad\ \vert\ q$
	$\Box q$
	$\Box p \supset \Box q$
	$\Box(p \supset q) \supset (\Box p \supset \Box q)$

R2 If $\vdash p$ then $\vdash \Box p$

If $\vdash p$, then there is a PC proof of p from A1–A3 and by our assumption there is an ND proof of p (which uses no premises). I.e.

$$\vdash \begin{array}{c} \cdot \\ \cdot \\ \cdot \\ p \end{array}$$

To obtain $\Box p$, simply put this within a *nec intro* subproof:

$$\Box \left\Vert\, \vdash \begin{array}{c} \cdot \\ \cdot \\ \cdot \\ p \end{array} \right.$$

$$\Box p$$

Since each formulation contains the other, they are equivalent.

APPENDIX III:

AN EFFECTIVE DECISION PROCEDURE

One of the disadvantages of the natural deduction formulations of logical systems, such as the one we have given, is that they do not provide an *effective* procedure for determining theoremhood. An *effective* procedure is a mechanical technique for determining in a finite number of steps whether or not a formula is a theorem. We know that if we can derive a formula with our rules it is valid, but if we find ourselves unable to come up with a proof on some occasion, we do not know whether that is due to a lack of ingenuity on our part or to the invalidity of the formula. Furthermore, our rules enable us to go in circles; we could just repeat lines or sequence of lines *ad infinitum* and never get to the end of a proof that could be finished. We are not required to use rules in a fixed order and not to repeat them indefinitely.

There are mechanical proof and disproof procedures for modal propositional logics, just as there are for classical propositional logic. The easiest ones to use are based on a technique developed by E. W. Beth, exploited by Kripke and Jeffrey and Zeman (q.v.). Procedures of this type for modal logic are developed in detail by Slaght, "Modal Tree Constructions," and Davidson, Jackson, and Parzetter, "Modal Trees for T and S5," in the *Notre Dame Journal of Formal Logic* Volume 18, No. 4, October, 1977, pp. 517–526 and pp. 602–605, respectively.

Tableau techniques are given for T, S4, and S5 and proven to be decision procedures in Zeman, J. Jay, *Modal Logic, The Lewis-Modal Systems*. Oxford: Clarendon Press, 1973, chaps. 9 and 14.

NOTES

1. For an excellent article presenting some central questions and main arguments, see Richard Cartwright's article "Propositions," in R. J. Butler, ed., *Analytical Philosophy,* First Series (Blackwell, 1962), pp. 81–103.

2. See Augustine's *On the Free Choice of the Will,* book III, chap. 3, and Boethius, *The Consolation of Philosophy,* book V, prose 6.

3. See Kretzmann, Kenny, and Pinborg, eds., *The Cambridge History of Later Medieval Philosophy* (Cambridge, 1982), p. 343. The translation is "loose" in the sense that I have rendered the modalities as applying only to propositions.

4. See B. Mates, *Stoic Logic* (University of California, 1953), chap. IV, §1.

5. See Anderson and Belnap, *Entailment* (Princeton, 1975).

6. E.g., by Lemmon in "New Foundations for Lewis Modal Systems," *The Journal of Symbolic Logic,* vol. 24 (1959), 1–14.

7. See Purtill, *Logic for Philosophers* (Harper & Row, 1971), chap. 7.3, and Purtill, "Four Valued Tables for Modal Logic," *Notre Dame Journal of Formal Logic,* vol. 11, no. 4, (October, 1970).

8. Even Anderson and Belnap find the implicational fragment of S4 acceptable as an account of "implies," although they do not accept the system as a whole.

9. This is a slight modification of a definition given by Perry, "Modalities in the *Survey* System of Strict Implication," *The Journal of Symbolic Logic,* vol. 4 (1939), 144, reported by Hughes and Cresswell, *An Introduction to Modal Logic* (Methuen, 1968), p. 50.

10. See Hughes and Cresswell, *Introduction to Modal Logic* (Methuen, 1968), p. 58, n. 37.

11. A. N. Prior, *Time and Modality* (Oxford, 1957), p. 49.

12. Ibid., and Prior and Fine, *Worlds, Times, and Selves* (Duckworth, 1977), pp. 102–3.

13. Lemmon, "Is There One Correct System of Modal Logic?" *The Aristotelian Society, Supp. Vol. XXXIII,* 1959, pp. 23–40. I draw heavily on Lemmon's article in what follows.

14. Ibid., p. 24.

15. Lewis and Langford, *Symbolic Logic* (Dover, 1959), p. 496.

16. See his *Time and Modality* (Oxford, 1957), *Past, Present, and Future* (Oxford, 1967), and *Papers on Time and Tense* (Oxford, 1968).

17. Hintikka, *Knowledge and Belief* (Cornell, 1962).

18. Lemmon mentions several more in the paper cited in note 13.

19. Those interested in pursuing this may consult Leonard Linsky, "Two Concepts of Quantification," *Nous*, vol. 6 (1972), 224–39. [Reprinted as an appendix to Linsky, *Names and Descriptions* (Chicago, 1977).] See also Ruth Barcan Marcus, "Interpreting Quantification," *Inquiry*, vol. 5 (1962), 252–59, and Nuel Belnap and J. Michael Dunn, "The Substitutional Interpretation of Quantifiers" *Nous*, vol. 2 (1968), 177–85; Susan Haack, *Philosophy of Logics* (Cambridge, 1978), chap. 4.

20. This technical term was defined a few pages earlier.

21. Quoted from Aristotle, *The Works of Aristotle*, ed. W. D. Ross, vol. I, (Oxford, 1928), 166a pp. 24–30.

22. *SCG*, I, 67. Tr. Anton Pegis (Notre Dame, 1975). Note that the "fallacy of composition and division" here mentioned is that of confusing the composite and the divided senses of some expression.

23. Several of these topics will arise again in "Some Objections to Quantified Modal Logic."

24. 'Divisible' itself appears to be a modal term. Here we will understand it to mean 'having at least two parts.'

25. Quine, "Three Grades of Modal Involvement," *The Ways of Paradox* (Random House, 1966), p. 174.

26. Quine, "Reference and Modality," in *From a Logical Point of View*, second edition (Harper and Row, 1961), p. 143.

27. Ibid., p. 144.

28. We will not pursue any further Quine's objections and possible responses. Anyone interested in pursuing the issue in detail should consult Leonard Linsky's anthology, *Reference and Modality* (Oxford, 1971), and the appendix to Alvin Plantinga's *The Nature of Necessity* (Oxford, 1974), where Quine's objection is discussed in detail.

29. And accidental properties as well. An accidental property is a property an individual has in the actual world but lacks in some other possible world in which the individual exists.

30. "Counterpart Theory and Quantified Modal Logic," *Journal of Philosophy*, vol. 65 (March 1968), pp. 113–26.

31. For more on fictional characters, see N. Wolterstorff's *Works and Worlds of Art* (Oxford, 1980), pp. 134–58.

32. What I give here is equivalent to the form $\Diamond(\exists x)ax \supset (\exists x)\Diamond ax$ which was first discussed by Ruth Barcan, now Ruth Barcan Marcus, in 1946.

33. So named by Alvin Plantinga, *The Nature of Necessity* (Oxford, 1974), p. 58, after Jean Buridan, who denied the truth of the natural interpretation of it.

34. "Quantification and Modality in S5," *Journal of Symbolic Logic*, vol. 21, no. 1 (March 1958), 60–62.

35. This objection is discussed in greater detail in chap. 7 of Plantinga's *The Nature of Necessity* and also in his "Actualism and Possible Worlds," *Theoria*, vol. 42 (1976), 139-60.

36. This section is based on Alvin Plantinga's "Actualism and Possible Worlds," *Theoria*, vol. 42 (1976), 139-60. The name "actualism" is from Robert M. Adams, "Theories of Actuality," *Nous*, vol. 8 (1974), 211-31. This "version" is an interpretation of quantified modal logic or an applied semantics.

37. These definitions come from Plantinga's *The Nature of Necessity*, chap. 5, and Thomas Jager's "An Actualistic Semantics for Quantified Modal Logic," *Notre Dame Journal of Formal Logic*, vol. 23, no. 3 (July 1982), 336-37.

38. At any rate, if one takes a realist view of properties, this will follow. Actualism in possible worlds obviously comports very well with metaphysical realism, even if it does not strictly require it.

39. "An Actualistic Semantics for Quantified Modal Logic," *Notre Dame Journal of Formal Logic*, vol. 23 (1982), 335-49.

40. Cf. *The Leibniz-Arnauld Correspondence*, ed. & tr., H. T. Mason (Manchester University Press and Barnes and Noble, 1967), pp. 15-16.

41. "Counterpart Theory and Quantified Modal Logic," *Journal of Philosophy*, vol. 65 (1968), pp. 113-26.

42. For a perspicuous statement of this problem, see Chisholm's "Identity through Possible Worlds; Some Questions," *Nous*, vol. 1 (1967), 1-8. Reprinted in Loux, *The Possible and the Actual* (Cornell, 1979).

BIBLIOGRAPHY

Ackrill, J. L. *Aristotle's Categories and De Interpretatione*. Clarendon Aristotle Series, ed. J. L. Ackrill. Oxford: Clarendon Press, 1963.

Adams, Robert M. "Theories of Actuality," *Nous,* vol. 8 (1974), 211-31.

Anderson, A. R., and Belnap, Nuel D., Jr. *Entailment: The Logic of Relevance and Necessity.* Princeton: Princeton University Press, 1975.

Aristotle, *De Sophisticis Elenchis,* in *The Works of Aristotle Translated Into English,* vol. I, W. D. Ross. Oxford: Clarendon Press, 1928. Newly reissued in *The Complete Works of Aristotle,* ed. Jonathan Barnes. Princeton: Princeton University Press, 1984.

Augustine. *On Free Will,* in *Augustine: Earlier Writings.* Tr. John H. S. Burleigh. Philadelphia: The Westminster Press, 1953.

Barcan, Ruth. "A Functional Calculus of First Order Based on Strict Implication," *Journal of Symbolic Logic,* vol. 11, (1946), 1-16.

Belnap, Nuel D., Jr., and J. Michael Dunn, "The Substitutional Interpretation of Quantifiers," *Nous,* vol. 2, (1968), 177-85.

Boethius, A. M. S., *The Consolation of Philosophy.* Tr. Richard Green. The Library of Liberal Arts. Indianapolis: The Bobbs Merrill Company, Inc., 1962.

Cartwright, Richard. "Propositions," in *Analytical Philosophy, First Series.* Ed. R. J. Butler. Oxford: Basil Blackwell, 1962. 81-103.

Chisholm, Roderick. "Identity through Possible Worlds: Some Questions," *Nous,* vol. 1 (1967), 1-8. Reprinted in Loux, *The Possible and the Actual* (Cornell, 1979).

Curley, E. M.
    1972   "Lewis and Entailment," *Philosophical Studies.* vol. 23, 198-204.
    1975   "The Development of Lewis' Theory of Strict Implication," *Notre Dame Journal of Formal Logic.* vol. 16, no. 4 (Oct. 1975), 517-27.

Davidson, B., F. C. Jackson, and R. Pargetter. "Modal Trees for T and S5," *Notre Dame Journal of Formal Logic,* vol. 18, no. 4 (October 1977), 602-07.

Fitch, Frederick B. *Symbolic Logic.* New York: Ronald Press, 1952.

Haack, Susan. *Philosophy of Logics.* Cambridge: Cambridge University Press, 1978.

Hintikka, Jaakko. *Knowledge and Belief.* Ithaca: Cornell University Press, 1962.

Hughes, G. E., and M. J. Cresswell. *An Introduction to Modal Logic.* London: Methuen, 1968.

Jager, Thomas. "An Actualistic Semantics for Quantified Modal Logic." *Notre Dame Journal of Formal Logic,* vol. 23, no. 3 (July 1982), 335–49.

Kretzmann, N., A. Kenny, and J. Pinborg, eds. *The Cambridge History of Later Medieval Philosophy.* Cambridge: Cambridge University Press, 1982.

Kripke, Saul.
    1959   "A Completeness Theorem in Modal Logic," *The Journal of Symbolic Logic.* vol. 24, no. 1 (March 1959), 1–14.
    1963a   "Semantical Considerations on Modal Logic," *Acta Philosophica Fennica.* Fasc. XVI, 83–94. Reprinted in Leonard Linsky, ed. *Reference and Modality.* Oxford: Oxford University Press, 1971.
    1963b   "Semantical Analysis of Modal Logic I: Normal Modal Propositional Calculi," *Zeitschrift fur mathematische Logik und Grundlagen der Mathematik.* vol. 9, 63–96.

Leibniz, G. W.
    1686   *Discourse on Metaphysics,* in *Philosophical Papers and Letters.* Second edition. Ed. Leroy E. Loemker. Dordrecht: Reidel Publishing Co., 1969.
    1686   *Leibniz-Arnauld Correspondence.* Tr. H. T. Mason. New York: Barnes & Noble, Inc., 1967.
    1714   *Monadology,* in *Philosophical Papers and Letters.* Second edition. Ed. Leroy E. Loemker. Dordrecht: Reidel Publishing Co., 1969.

Lemmon, E. J.
    1957   "New Foundations for Lewis Modal Systems," *The Journal of Symbolic Logic.* vol. 22 (June 1957), 176–86.
    1959   "Is There Only One Correct System of Modal Logic?" *The Aristotelian Society, Supplementary Volume XXXIII,* 1959. London: Harrison and Son, Ltd., 23–40.

Lewis, C. I., and C. H. Langford. *Symbolic Logic.* Second Edition. New York: Dover Publications, 1959.

Lewis, David. "Counterpart Theory and Quantified Modal Logic." *Journal of Philosophy.* vol. 65 (March 1968), 113–28.

Linsky, Leonard.
    1971   Ed. *Reference and Modality.* Oxford: Oxford University Press.
    1972   "Two Concepts of Quantification." *Nous.* vol. 6, 224–39. Reprinted as an appendix to Linsky, *Names and Descriptions,* Chicago: University of Chicago Press, 1977.

Loux, Michael, ed. *The Possible and the Actual.* Ithaca: Cornell University Press, 1979.

Marcus, Ruth Barcan (see also Barcan, Ruth). "Interpreting Quantification." *Inquiry,* vol. 5 (1962), 252–59.

Mates, Benson. *Stoic Logic.* University of California Publications in Philosophy, vol. 26. Berkeley and Los Angeles: University of California Press, 1953.

Mendelson, Elliot. *Introduction to Mathematical Logic.* Second edition. New York: Van Nostrand, 1979.

Plantinga, Alvin.

1974   *The Nature of Necessity.* Oxford: Clarendon Press.

1976   "Actualism and Possible Worlds." *Theoria,* vol. 42, 139-60. Reprinted in M. Loux, *The Possible and the Actual,* Ithaca: Cornell University Press, 1979.

Prior, Arthur Norman.

1955   *Formal Logic.* Oxford: Clarendon Press.

1956   "Modality and Quantification in S5," *The Journal of Symbolic Logic,* vol. 21, 60-62.

1957   *Time and Modality.* Oxford: Clarendon Press.

1967   *Past, Present, and Future.* Oxford: Clarendon Press.

Prior, Arthur Norman, and Kit Fine. *Worlds, Times, and Selves.* Amherst: University of Massachusetts Press and London: Duckworth, 1977.

Purtill, Richard.

1970   "Four-Valued Tables for Modal Logic," *Notre Dame Journal of Formal Logic,* vol. 11, no. 4 (October 1970).

1971   *Logic for Philosophers.* New York: Harper and Row.

Quine, Willard Van Orman.

1961   *From a Logical Point of View.* Second edition. Harper Torchbook Edition. New York: Harper and Row.

1966   *The Ways of Paradox and Other Essays.* New York: Random House.

Slaght, Ralph L. "Modal Tree Constructions," *Notre Dame Journal of Formal Logic.* vol. 18, no. 4 (October 1977), 519-26.

Thomas Aquinas, St. *Summa Contra Gentiles.* Tr. Anton C. Pegis. Notre Dame: University of Notre Dame Press, 1975. Reprint of *On the Truth of the Catholic Faith.* Garden City, N.Y.: Hanover House, 1955.

Whitehead, Alfred North, and Bertrand Russell. *Principia Mathematica to *56.* Cambridge: Cambridge University Press, 1962.

Zeman, J. Jay. *Modal Logic: The Lewis-Modal Systems.* Oxford: Clarendon Press, 1973.